Sonochemical Reactions

Edited by Selcan Karakuş

Published in London, United Kingdom

IntechOpen

Supporting open minds since 2005

Sonochemical Reactions
http://dx.doi.org/10.5772/intechopen.83195
Edited by Selcan Karakuş

Contributors
Raed Al-Juboori, Les Bowtell, Dang Van Thanh, Van-Truong Nguyen, Nhat Huy Nguyen, Pham Van Hao, Quoc Dung Nguyen, Prashant Devidas Jolhe, Bharat Bhanavase, Satish Mardikar, Vilas Patil, Shirish Sonawane, Lingayya Hiremath, Jitendra Pal Singh, Ganesh Pandey, Manish Kumar, Aditya Sharma, Xingrui Chen, Qichi Le, Ignacio Alvarez, Leire Astráin-Redín, Guillermo Cebrián, Javier Raso, Santiago Condón, Salomé Ciudad-Hidalgo, Selcan Karakuş, Ezgi Tan, Merve Ilgar, Ibrahim Mizan Kahyaoglu, Yeşim Müge Sahin, Demet Sezgin Mansuroglu, Deniz Ismik, Nevin Tasaltin, Ayben Kilislioglu

First published in London, United Kingdom, 2020 by IntechOpen
IntechOpen is the global imprint of INTECHOPEN LIMITED, registered in England and Wales, registration number: 11086078, 7th floor, 10 Lower Thames Street, London, EC3R 6AF, United Kingdom
Printed in Croatia

British Library Cataloguing-in-Publication Data
A catalogue record for this book is available from the British Library

Additional hard and PDF copies can be obtained from orders@intechopen.com

Sonochemical Reactions
Edited by Selcan Karakuş
p. cm.
Print ISBN 978-1-83880-001-7
Online ISBN 978-1-83880-002-4
eBook (PDF) ISBN 978-1-83880-211-0

For EU product safety concerns:
IN TECH d.o.o., Prolaz Marije Krucifikse Kozulić 3, 51000 Rijeka, Croatia, info@intechopen.com or visit our website at intechopen.com.

Meet the editor

Assistant Professor Selcan Karakuş is currently working in the Department of Chemistry, Istanbul University-Cerrahpasa (IUC), Turkey. She received her Master of Science degree in Physical Chemistry from Istanbul University (IU) in 2006. She received her Doctor of Philosophy degree in Physical Chemistry from IU in 2011. She worked as a visiting researcher at the University of Massachusetts, Department of Polymer Science and Engineering. She has research experience in drug carrier systems, nanoparticles, nanocomposites, nanoemulsion self-assembled polymeric nanostructures, and copolymer blends. She has worked on different projects funded by Istanbul University-Cerrahpasa. She has published several research articles and book chapters in this area.

Contents

Preface

The sonochemistry method is preferred in different application areas (biomedical, textile, dye, and environmental) to prepare nano-sized materials. This book, Sonochemical Reactions, comprises eight chapters that provide an overview to the definition of the sonochemical method and applications areas (biotechnolgical, food, environmental).

The aim of the sonochemical process is to produce high energy to uniformly disperse a solution. In the ultrasound-assited method, the formation, growth, and explosive collapse of bubbles in a liquid occurs due to acoustic cavitation. This method is known as green synthesis in the production of small sized materials.

Recently, in the preparation of nanostructures, the sonochemical method has been preferred because it is economical, simple, and does not require high temperature and energy. Occupational health and safety, economy, and time are the most important factors in the synthesis of superior nanostructures. In the sonochemical method, in the case of cavitation, intense heating (5000°C) and pressure (500 atm) exposure of the bubbles occurs in a short time period (microseconds).

It is emphasized in this book that aqueous systems modified by ultrasonic irradiation behave differently according to their physical and chemical properties.

The sonochemical degradation of chemicals was investigated in various environmental samples (sea water, urban runoff, and domestic wastewater) as well as in model solutions (pure and salt water).

The synthesis of different nanostructures such as polymeric nanoparticles and metal oxide nanoparticles were exemplified in detail in this book. The role of sonochemistry in the synthesis of graphene-based nanosheets, which are a trend material in recent days, is explained.

With the support of the authors of this book, this book has added a new critical value to different fields such as ultrasonic assisted adsorption and food application.

We sincerely thank our authors who have contributed with experience and knowledge to this book. Especially, our thanks go to Haydar Düven, Serife Düven, Alpaslan Düven, and the editorial team from IntechOpen Publishing for their assistance in preparing this book.

Selcan Karakuş
Assistant Professor,
Istanbul University-Cerrahpaşa,
Faculty of Engineering,
Department of Chemistry,
Istanbul, Turkey

Section 1

Ultrasound-Assisted Synthesis

Chapter 1

Preparation, Characterization, and Swelling Behavior of PEGylated Guar Gum @ Ag Nanoparticles

Selcan Karakus, Ezgi Tan, Merve Ilgar, Ibrahim Mizan Kahyaoglu, Yeşim Müge Şahin, Demet Sezgin Mansuroglu, Deniz Ismik, Nevin Tasaltin and Ayben Kilislioglu

Abstract

In this study, polyethylene glycol/guar gum @ silver nanoparticles (PEG/GG@AgNPs) were synthesized by using simple sonication method. The nanoparticles were characterized using Fourier-transform infrared spectroscopy (FTIR) and scanning transmission electron microscopy (STEM). The swelling behaviors of nanoparticles were studied in different pHs (5.5 and 7.4). The experimental results were calculated by Fickian diffusion and Schott kinetic models to understand the swelling mechanism and coefficients of the nanoparticles. The results showed that the linear equation of the Fickian diffusion kinetic model was best fit to explain the water diffusion mechanism of the nanoparticle with high correlation coefficient (R^2 = 0.982–0.987). The results confirmed that the swelling degree of nanoparticles were 9.71 g/g at pH 5.5. Also, the results confirmed that PEG/GG@AgNPs can be a good candidate for drug delivery systems in pharmaceutical applications.

Keywords: swelling behavior, guar gum, Ag nanoparticles

1. Introduction

Nanotechnology focuses on many fundamental disciplines such as physics, chemistry, materials science, and biology [1]. Recently, the synthesis of the nanostructures has gained a great attention due to superior properties (mechanical, optical, thermal, biological, physical, and chemical) as compared to conventional materials [2–5].These superior properties depend on the size, composition, shape, and origin of nanostructures [6, 7]. As a general definition, nanostructures are 1–100 nanometers in size in which case they have a high surface area-to-volume ratios and their reactivities are effected mostly depending on their different shapes such as spherical, conical, spiral, cylindrical, tubular, and hollow [8, 9].

Many researchers have reported that iron [10], copper [11], gold [2], and silver [12] were used often for the preparation of stable dispersions of nanoparticles due to their biocompatibility and biodegradability and less reactivity in the biomedical applications. Silver nanoparticles (AgNPs) with desired morphologies, well-known for their antimicrobial activity, are used in both ionic and metallic forms which are incorporated inside the polymer matrix, with excellent biocompatibility [13]. It was known that bare silver nanoparticles were prone to oxidation and tarnishing. Thus, we used PEG/GG polymer blends to improve the dispersion stability and prevent agglomeration of silver nanoparticles in aqueous environment [14].

During the preparation of hydrogels, mostly preferred biopolymers are alginate, chitosan, gum arabic, agar, carrageenans, and guar gum [15–19]. Among these biopolymers, GG is used in the fields of food industry and pharmaceutical and cosmetic applications [20–22]. It is not desirable to prepare silver nanoparticles by chemical methods using toxicological chemicals such as reducing agents [23]. For this purpose, we chose a dual biopolymer blend (GG and PEG) which will carry out the green synthesis and improve the distribution. GG is a nonionic water-soluble polysaccharide and consists of galactomannan which has a linear chain of (1-4)-β-D-mannopyranosyl units interposed with (1-6)-α-D-galactopyranosyl units distributed as side branches [24]. PEG is a stabilizer effective for the control of size and shape of nanoparticles and also has role on the reduction of silver ions [25].

The key points of the chapter were (i) the green synthesis and (ii) swelling of nanoparticles. The aim of this study was to obtain NPs consisting of PEG/GG and zero valent Ag by using the simple sonication method. We prepared nanoparticles in the presence of silver nitrate in GG/PEG (2:1 mixing ratio) polymer blends. In particular, we performed a green synthesis sonochemical process to reduce Ag ions to form AgNPs by using ultrasonic method without the use of dangerous stabilizing agents such as any reducing agent and surfactant [26]. The breaking of cavitation bubbles under high temperature and pressure and the hydrogen radicals (H^*) and hydroxyl radicals (OH^*) in the water formed by the ultrasonic effect play an important role in the reduction of Ag ions and the formation of AgNPs [27].

All samples were characterized by using FTIR and STEM techniques. SEM images revealed good compatibility and homogeneous distribution between the PEG/GG matrix and Ag. AgNPs were found to be <500 nm in size. Moreover, we have also demonstrated the swelling behavior of the prepared nanoparticles in finding out the potential of the nanoparticles for drug delivery systems. The swelling uptake (%) of PEG/GG@AgNPs was found to be % 670.5 at pH 5.5.We determined the swelling behavior such as the maximum swelling and gel fraction of samples to interpret the water absorption results. All results showed that one could control the size and the shape of zero valent Ag nanoparticles by polymer blend under the sonication effect.

2. Materials and methods

2.1 Materials

Guar gum (99% purity, average molecular weight of 2.8×10^5 g mol^{-1}) and polyethylene glycol (PEG 400) were obtained from Fluka (Switzerland). Silver nitrate ($AgNO_3$), sodium hydroxide (NaOH), and dimethyl sulfoxide (DMSO) were obtained from Merck (Pvt.) Ltd. Mumbai, India. Polyethylene glycol 400 (PEG) (molecular weight of 400 g/mol) was purchased from Sigma-Aldrich, Chemie GmbH, USA.

2.2 Preparation of nanoparticles

2.2.1 Preparation of PEG/GG@AgNPs

GG solution (0.5 g GG in 100 mL of deionized water) and PEG (5 mL) were mixed by stirring with a magnetic stirrer at 25°C for 10 min. 5 mL of $AgNO_3$ (0.1 M) solution was added into GG/PEG solutions and then was sonicated for 15 min. 0.1 M NaOH was added into the solutions until pH 8.4.

2.3 Characterization

Samples were ground with KBr powder and analyzed from 4000 to 600 cm^{-1} with a resolution of 4 cm^{-1} using eight scans by using a PerkinElmer FTIR emission spectrometer (Spectrum Two). Samples were scanned in the dark-field area with the wet STEM detector by using FEI QUANTA S50 (A copper grid, Ted Pella, support films, carbon type A, 300 meshes was utilized). STEM holder was cooled to 2°C and the pressure was set between 700 and 1300 Pa.

2.4 Swelling studies

2.4.1 Measurement of the water absorbency

The swelling degree (St, g/g) was calculated from Eq. (2):

$$S_t = \frac{W_t - W_d}{W_d} \tag{1}$$

where S_W is the swelling degree per gram dried sample ($g\ g^{-1}$), W_d is the mass of dried samples at time t (g), and W_t is the mass of swollen samples at time t (g) [28].

2.4.2 Calculations of the swelling behavior

The swelling behavior was explained by applying different kinetic models such as Fickian diffusion and Schott second-order dynamic model given in Eqs. (2)–(7) [29, 30]. To identify the swelling kinetic mechanisms of the prepared samples, the swelling kinetic parameters were evaluated according to models:

Fickian diffusion swelling kinetic models:

$$S_t = S_\infty x\left(1 - e^{-k_W t}\right) \tag{2}$$

When Eq. (2) was linearized

$$Ln\frac{S_t}{S_\infty} = Lnk + nLnt \tag{3}$$

where t is time (min), k is the rate constant (min^{-1}), St is the water absorption capacity at time t, and S_∞ is the water absorption capacity at equilibrium. The plots of LnS_t/S_∞ versus Ln t were drawn to calculate the parameter (k) of kinetic model and the linear correlation coefficient.

Schott proposed the second-order kinetic model to elucidate the swelling mechanism of the system, and this model was related to the swelling rate at any time and was proportional to the quadratic of the swelling capacity before the equilibrium state [30, 31]:

$$\frac{dS}{dt} = k_s(S_t - S_\infty)^2 \quad (4)$$

When the initial condition was t = 0 and S = 0

$$\frac{t}{S} = A + Bt \quad (5)$$

$$A = 1/_{k_s S_\infty^2} \quad (6)$$

where t is time (min), S is the swelling capacity at time t (g g^{-1}), A is the reciprocal of initial swelling rate, and B is the reciprocal of S_∞. The plots of t/W versus t were drawn to calculate the parameters (A and B) of kinetic model and the linear correlation coefficient.

To explain the water diffusion mechanism, fractional solution capacity (F: S_t/ S_∞) less than 0.60 was analyzed using the following equation (Eq. (7)):

$$\frac{S_t}{S_\infty} = kt^n \text{ or } \quad F = kt^n \quad (7)$$

where t is time (min), Wt is the uptake capacity of the sample at time t, S_∞ is the capacity of the sample at swelling equilibrium (g), k is the gel characteristic constant, and n is the swelling index (n < 0.5 for Fickian diffusion and 0.5 < n < 1 for non-Fickian; n = 1 for relaxation controlled transport and n > 1 for II diffusion) [30].

The Fickian diffusion model had another expression shown as Eq. (8):

$$\frac{S_t}{S_\infty} = 4\sqrt{\left(D/_{\pi l^2}\right)}\left[\frac{1}{\pi^2} + 2\sum_{n=1}^{x}(-1)^n ierfc\left(\frac{nl}{2\sqrt{Dt}}\right)\right] \quad (8)$$

Diffusion coefficient D (cm^2/s) was calculated using Eq. (9). For short times, Eq. (9) was used at the early-stage diffusion coefficient of water by

$$\frac{S_t}{S_\infty} = 4\sqrt{\left(D/_{\pi l^2}\right)} \quad (9)$$

where l was the diffusional distance [32, 33].

3. Results and discussions

3.1 Swelling kinetics

Swelling ability is known to be associated with free hydrophilic groups and surface properties [34]. The sonication method, which is associated with various factors such as ultrasound power, ultrasound frequency, modification time, and temperature, changes the surface properties [35]. In this study, the swelling kinetics was investigated to determine the surface properties of the novel PEG/GG@AgNPs obtained by using sonication method. To analyze the effect of sonication on the swelling kinetic mechanism of the synthesized nanoparticles, swelling kinetic results were observed gravimetrically and performed in pH 5.5 and 7.4 at 25°C.

The equilibrium swelling degrees, Fickian and Schott kinetic models of PEG/GG@AgNPs were given in **Figures 1–3**, respectively and the swelling behavior followed a similar behavior due the high surface area.

The O-H groups of the PEG/GG@AgNPs formed hydrogen bonds with water molecules and absorbed water [36]. The equilibrium swelling degrees of PEG/GG@AgNPs were 9.71 gg^{-1} (pH 5.5) and 3.82 gg^{-1} (pH 7.4). According to the experimental swelling results, the swelling degree of the PEG/GG@AgNPs increased within the first 10 minutes and then slowed down until reaching equilibrium after 15 min in two different pH mediums. In this case, the results showed that the nanoparticle had shown greater interest in the water molecules and was proof

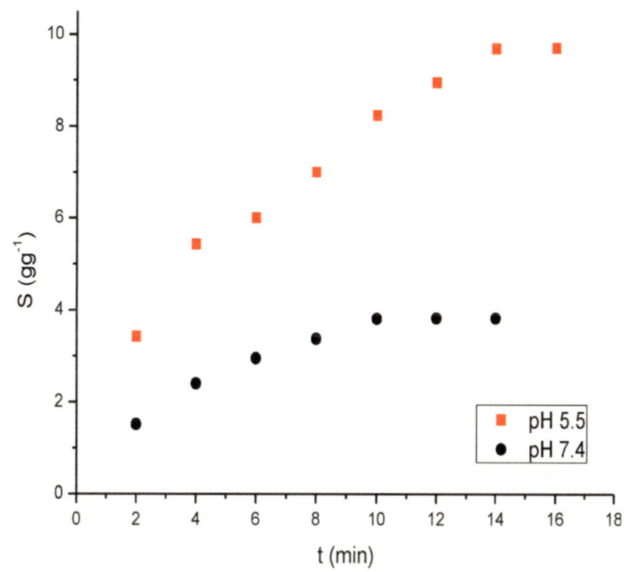

Figure 1.
Effect of two different pH mediums on the swelling kinetics of PEG/GG@AgNPs.

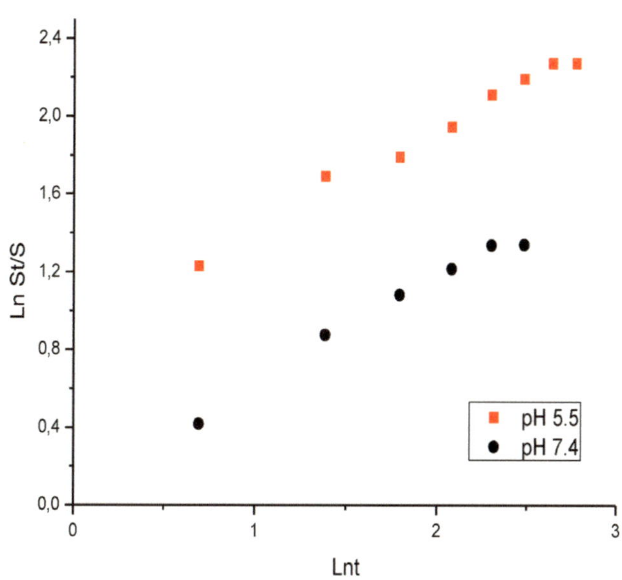

Figure 2.
Plots of Fickian kinetic model for PEG/GG@AgNPs.

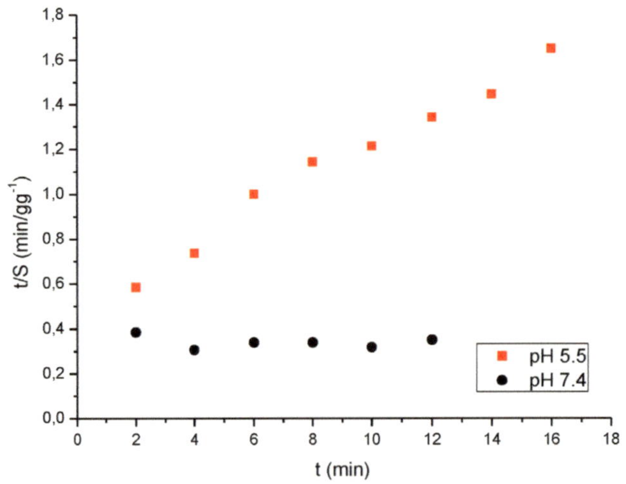

Figure 3.
Plots of Schott kinetic model for PEG/GG@AgNPs.

Sample	Fickian kinetic model			Schott kinetic model		
	n	k	R^2	$S_\infty(exp)$	k_2 (g/g min^{-1})	R^2
pH 5.5	0.51	2.50	0.987	1.39	1.052	0.978
pH 7.4	0.53	2.72	0.982	52.63	0.001	0.070

Table 1.
Values of various parameters associated with swelling kinetic models.

that it binds with O—H groups in the structure which allowed it to swell in a short time in pH 5.5. In order to explain swelling behavior of the nanoparticle, the swelling parameters were calculated by using the Fickian diffusion and Schott's second-order kinetic models. The calculated swelling kinetic parameters and the correlation coefficients (R^2) for all the models for the nanoparticles were given in **Table 1**.

The results showed that the linear equation of the Fickian diffusion kinetic model was best fit to explain the water diffusion mechanism of the nanoparticles with high correlation coefficient (R^2 = 0.982–0.987). According to the Fickian kinetic model, n is known to explain the diffusion mechanism of the solvent. The n values of the nanoparticles were calculated to be in the range of 0.5–1, which was explained by a non-Fickian diffusion behavior of the water transport mechanism [36].

3.2 STEM analysis

To investigate the effect of ultrasonic irradiation on the surface properties of the nanoparticle, the STEM images of PEG/GG@AgNPs were presented in **Figure 4**. It could be clearly seen that the surface of PEG was a homogeneous surface in spherical nanoform with the uniform dispersion of AgNPs. It was concluded that the ultrasonic irradiation may play a role in obtaining homogeneous distribution of the nanoparticles.

With this structural feature, nanoparticles have been used in different applications such as sensor, drug delivery system, and pharmaceutical applications [37–41].

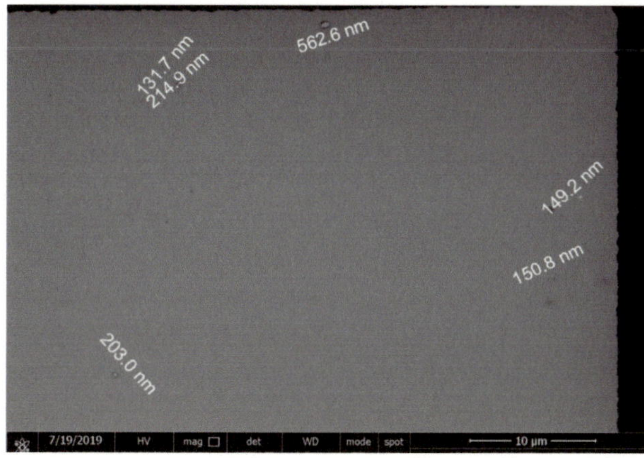

Figure 4.
STEM image of PEG/GG@AgNPs.

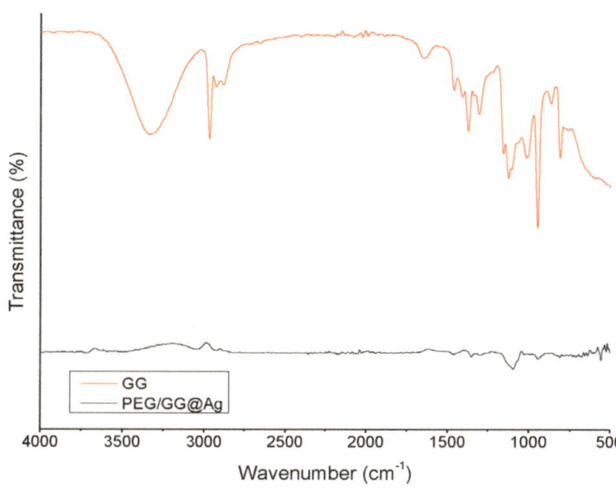

Figure 5.
The FTIR spectrums of pure GG and PEG/GG@AgNPs.

3.3 FTIR analysis

The FTIR spectrums of pure GG and PEG/GG@AgNPs were given in **Figure 5**. The FTIR spectrum of pure GG showed peaks at 3499 (—OH stretching), 2989 (C—H stretching), and 1030 (—OH bending) cm^{-1}. The FTIR spectrum of PEG/GG@AgNPs showed peaks at 2980 (C—H stretching), 1540 (COOH), 1470 (C—H stretching), and 1370 (—C=O) cm^{-1}. The peaks recorded at 3499 cm^{-1} of GG expressed in AgNPs indicated that -OH groups were utilized for the stabilization of AgNPs. From the results of the FTIR, we found that reduction and stabilization occurred in AgNPs. Singh et al. observed similar results [42].

4. Conclusions

In this study, PEG/GG@AgNPs were prepared by using sonochemical method. The swelling ability of nanoparticle was investigated in two different pHs. The mechanism of swelling kinetics was explained, and it has been found that the mechanism follows the Fickian diffusion model. In summary, this study was focused on the green, low-cost novel method for producing Ag nanoparticles. The NPs could be served as a promising candidate nanocarrier for drug delivery systems due to its the swelling degrees.

Acknowledgements

The authors acknowledge the STEM and FTIR analysis support from ArelPOTKAM, Istanbul, and Zeynep Akça.

Author details

Selcan Karakus[1*], Ezgi Tan[1], Merve Ilgar[1], Ibrahim Mizan Kahyaoglu[1], Yeşim Müge Şahin[2,3], Demet Sezgin Mansuroglu[2,4], Deniz Ismik[2,5], Nevin Tasaltin[6] and Ayben Kilislioglu[1]

1 Faculty of Engineering, Department of Chemistry, Istanbul University-Cerrahpasa, Istanbul, Turkey

2 ArelPOTKAM (Polymer Technologies and Composite Application and Research Center), Istanbul Arel University, Istanbul, Turkey

3 Faculty of Engineering and Architecture, Department of Biomedical Engineering, Istanbul Arel University, Istanbul, Turkey

4 Department of Chemistry, Yildiz Technical University, Davutpasa Campus, Istanbul, Turkey

5 Faculty of Chemistry and Metallurgy, Department of Bioengineering, Yildiz Technical University, Istanbul, Turkey

6 Department of Electrical-Electronics Engineering, Maltepe University, Istanbul, Turkey

*Address all correspondence to: selcan@istanbul.edu.tr

References

[1] Bensaude-Vincent B. Building multidisciplinary research fields: The cases of materials science, nanotechnology and synthetic biology. In: The Local Configuration of New Research Fields. Cham: Springer; 2016. pp. 45-60

[2] Daniel MC, Astruc D. Gold nanoparticles: Assembly, supramolecular chemistry, quantum-size-related properties, and applications toward biology, catalysis, and nanotechnology. Chemical Reviews. 2004;**104**(1):293-346

[3] Abbasi E, Milani M, Fekri Aval S, Kouhi M, Akbarzadeh A, Tayefi Nasrabadi H, et al. Silver nanoparticles: Synthesis methods, bio-applications and properties. Critical Reviews in Microbiology. 2016;**42**(2):173-180

[4] Bang JH, Suslick KS. Applications of ultrasound to the synthesis of nanostructured materials. Advanced Materials. 2010;**22**(10): 1039-1059

[5] Wang ZL. Functional oxide nanobelts: Materials, properties and potential applications in nanosystems and biotechnology. Annual Review of Physical Chemistry. 2004;**55**: 159-196

[6] Scott BJ, Wirnsberger G, Stucky GD. Mesoporous and mesostructured materials for optical applications. Chemistry of Materials. 2001;**13**(10): 3140-3150

[7] Yin Y, Talapin D. The chemistry of functional nanomaterials. Chemical Society Reviews. 2013;**42**(7):2484-2487

[8] Sun X. Morphology and size-controllable preparation of silver nanostructures through a wet-chemical route at room temperature. Inorganic Materials. 2010;**46**(6):679-682

[9] Rosi NL, Mirkin CA. Nanostructures in biodiagnostics. Chemical Reviews. 2005;**105**(4):1547-1562

[10] Wu W, He Q, Jiang C. Magnetic iron oxide nanoparticles: Synthesis and surface functionalization strategies. Nanoscale Research Letters. 2008; **3**(11):397

[11] Eastman JA, Choi SUS, Li S, Yu W, Thompson LJ. Anomalously increased effective thermal conductivities of ethylene glycol-based nanofluids containing copper nanoparticles. Applied Physics Letters. 2001;**78**(6): 718-720

[12] Rai M, Yadav A, Gade A. Silver nanoparticles as a new generation of antimicrobials. Biotechnology Advances. 2009;**27**(1):76-83

[13] Hasan A, Waibhaw G, Saxena V, Pandey LM. Nano-biocomposite scaffolds of chitosan, carboxymethyl cellulose and silver nanoparticle modified cellulose nanowhiskers for bone tissue engineering applications. International Journal of Biological Macromolecules. 2018;**111**:923-934

[14] Abdullah MF, Ghosh SK, Basu S, Mukherjee A. Cationic guar gum orchestrated environmental synthesis for silver nano-bio-composite films. Carbohydrate Polymers. 2015;**134**:30-37

[15] Nair LS, Laurencin CT. Biodegradable polymers as biomaterials. Progress in Polymer Science. 2007;**32**(8-9):762-798

[16] Kumar T, Gupta SK, Prajapati MK, Tripathi D. Natural excipients: A review. Asian Journal of Pharmacy and Life Science. 2012;**2**(1):97-108. ISSN 2231–4423

[17] Anal AK. Bionanotechnology: Principles and Applications. USA: CRC Press; 2018

[18] Makshakova ON, Faizullin DA, Zuev YuF. Interplay between secondary structure and ion binding upon thermoreversible gelation of κ-carrageenan. Carbohydrate Polymers. 2020;**227**:115342

[19] Tiwari P, Panthari P, Katare DP, Kharkwal H. Natural polymers in drug delivery. World Journal of Pharmaceutical Sciences. 2014;**3**(9): 1395-1409

[20] Mudgil D, Barak S, Khatkar BS. Guar gum: Processing, properties and food applications—A review. Journal of Food Science and Technology. 2014; **51**(3):409-418

[21] Jani GK, Shah DP, Prajapati VD, Jain VC. Gums and mucilages: Versatile excipients for pharmaceutical formulations. Asian Journal of Pharmaceutical Sciences. 2009;**4**(5): 309-323

[22] Parente ME, Ochoa Andrade A, Ares G, Russo F, Jiménez-Kairuz Á. Bioadhesive hydrogels for cosmetic applications. International Journal of Cosmetic Science. 2015;**37**(5):511-518

[23] Kanmani P, Rhim JW. Physicochemical properties of gelatin/silver nanoparticle antimicrobial composite films. Food Chemistry. 2014; **148**:162-169

[24] Palem RR, Madhusudana Rao K, Kang TJ. Self-healable and dual-functional guar gum-grafted-polyacrylamidoglycolic acid-based hydrogels with nano-silver for wound dressings. Carbohydrate Polymers. 2019;**223**:115074

[25] Shameli K, Bin Ahmad M, Jazayeri SD, Sedaghat S, Shabanzadeh P, Jahangirian H, et al. Synthesis and characterization of polyethylene glycol mediated silver nanoparticles by the green method. International Journal of Molecular Sciences. 2012;**13**(6):6639-6650

[26] Kora AJ, Sashidhar RB, Arunachalam J. Gum kondagogu (Cochlospermum gossypium): A template for the green synthesis and stabilization of silver nanoparticles with antibacterial application. Carbohydrate Polymers. 2010;**82**(3):670-679

[27] He C, Liu L, Fang Z, Li J, Guo J, Wei J. Formation and characterization of silver nanoparticles in aqueous solution via ultrasonic irradiation. Ultrasonics Sonochemistry. 2014;**21**(2): 542-548

[28] Wang Y, Xiong Y, Wang J, Zhang X. Ultrasonic-assisted fabrication of montmorillonite-lignin hybrid hydrogel: Highly efficient swelling behaviors and super-sorbent for dye removal from wastewater. Colloids and Surfaces A: Physicochemical and Engineering Aspects. 2017;**520**:903-913

[29] Wang X, Hou H, Li Y, Wang Y, Hao C, Ge C. A novel semi-IPN hydrogel: Preparation, swelling properties and adsorption studies of Co (II). Journal of Industrial and Engineering Chemistry. 2016;**41**:82-90

[30] Gharekhani H, Olad A, Mirmohseni A, Bybordi A. Superabsorbent hydrogel made of NaAlg-g-poly (AA-co-AAm) and rice husk ash: Synthesis, characterization, and swelling kinetic studies. Carbohydrate Polymers. 2017;**168**:1-13

[31] Wang M, Xu L, Ju X, Peng J, Zhai M, Li J, et al. Enhanced radiation crosslinking of carboxymethylated chitosan in the presence of acids or polyfunctional monomers. Polymer Degradation and Stability. 2008;**93**(10):1807-1813

[32] Vuković JS, Babić MM, Antić KM, Miljković MG, Perić-Grujić AA, Filipović JM, et al. A high efficacy antimicrobial acrylate based hydrogels with incorporated copper for wound healing application. Materials Chemistry and Physics. 2015;**164**:51-62

[33] Tanc B, Orakdogen N. Charged groups synergically enhanced elasticity and tunable swelling/shrinking of poly (dialkylaminoethyl methacrylate)/ layered silicate nanocomposite cryogels. Polymer. 2019;**178**:121627

[34] Lee JN, Park C, Whitesides GM. Solvent compatibility of poly (dimethylsiloxane)-based microfluidic devices. Analytical Chemistry. 2003; **75**(23):6544-6554

[35] Jamalabadi M, Saremnezhad S, Bahrami A, Jafari SM. The influence of bath and probe sonication on the physicochemical and microstructural properties of wheat starch. Food Science & Nutrition. 2019;**7**(7):2427-2435

[36] Dai H, Zhang H, Ma L, Zhou H, Yu Y, Guo T, et al. Green pH/magnetic sensitive hydrogels based on pineapple peel cellulose and polyvinyl alcohol: Synthesis, characterization and naringin prolonged release. Carbohydrate Polymers. 2019;**209**:51-61

[37] Slowing II, Trewyn BG, Giri S, Lin VY. Mesoporous silica nanoparticles for drug delivery and biosensing applications. Advanced Functional Materials. 2007;**17**(8):1225-1236

[38] West JL, Halas NJ. Engineered nanomaterials for biophotonics applications: Improving sensing, imaging, and therapeutics. Annual Review of Biomedical Engineering. 2003;**5**(1):285-292

[39] Soomro RA, Kalwar NH, Avci A, Pehlivan E, Hallam KR, Willander M. In-situ growth of NiWO4 saw-blade-like nanostructures and their application in photo-electrochemical (PEC) immunosensor systems designed for the detection of neuron-specific enolase. Biosensors and Bioelectronics. 2019;**141**: 111331

[40] Tunesi MM, Kalwar NH, Soomro RA, Karakus S, Jawaid S, Abro MI. Tartaric acid assisted in-situ growth of CuO nanostructures over ITO substrate for the electrocatalytic detection of Sudan I. Materials Science in Semiconductor Processing. 2018;**75**: 296-300

[41] Tunesi MM, Kalwar N, Abbas MW, Karakus S, Soomro RA, Kilislioglu A, et al. Functionalised CuO nanostructures for the detection of organophosphorus pesticides: A non-enzymatic inhibition approach coupled with nano-scale electrode engineering to improve electrode sensitivity. Sensors and Actuators B: Chemical. 2018;**260**: 480-489

[42] Singh J, Dhaliwal AS. Synthesis, characterization and swelling behavior of silver nanoparticles containing superabsorbent based on grafted copolymer of polyacrylic acid/Guar gum. Vacuum. 2018;**157**:51-60

Chapter 2

Bottom-Up and Top-Down Approaches for MgO

Jitendra Pal Singh, Manish Kumar, Aditya Sharma, Ganesh Pandey, Keun Hwa Chae and Sangsul Lee

Abstract

In this chapter, we present an overview of synthesis of MgO nanoparticles and thin films by using top-down and bottom-up approaches. The bottom-up approaches are generally utilized to grow nanoparticles by the methods that involve chemical reactions. Sometimes, methods based on these reactions are also able to grow thin films. The top-down approaches are preferred for growing thin films where bulk material is used for depositions. The methods, which are frequently used, are radio frequency sputtering, pulsed lased deposition, and molecular beam epitaxy and e-beam evaporation. Sometimes, methods like mechanical milling and high energy ball milling are used to grow nanoparticles.

Keywords: MgO, bottom-up approaches, top-down approaches

1. Introduction

Nanoparticles and thin films are very common form of materials for utilization in different applications [1–4]. Synthesis approaches play vital role to determine characteristics of nanoparticles [5] and thin films [6]. Thus, a number of methods are being developed to synthesize either nanoparticles [7–9] or thin films [10–12]. The motive behind to explore numerous methods is to look for reproducibility and cost effectiveness in terms of industrial utilization [13, 14]. Researchers are also working to get deep insights of involved phenomena during growth which persists a way to optimize for particular application [15–18]. The factors, which are considered during nanoparticle growth, are size [19], shape [20, 21] and size distribution [22, 23]. In case of thin films, these factors are nature of growth, morphology, stress, strain developed across films substrate interface [24–26].

While growing nanoparticles, one need to take care annealing treatment [27, 28] and stoichiometry [29, 30], however, process is rather typical in case of thin film technology. Choice of substrate [31], annealing temperature [32, 33], base pressure [34], target to substrate distance [35], deposition pressure [36, 37] and nature of gas during growth determine the nature of film [38]. Textured of grown thin film [39], stoichiometry [40] and nature of surface [41, 42] are another important parameter, which are considered during deposition. Thus, keeping in mind the necessity and challenges in the synthesis, synthesis approaches for growing nanoparticles and thin films are discussed by taking a simple inorganic system. However, magnesium oxide is known from long time [43] but recent advances in application of this material motivated us to discuss these approaches for MgO [44]. In **Table 1**, a summary of properties of MgO are depicted [45–47].

While keeping in mind the importance of this material, we attempt to give an overview of synthesis of MgO nanoparticle and thin film. To grow nanoparticles, two kinds of approaches are used: (1) bottom-up approach and (2) top-down approach [48, 49]. These approaches are explained on the basis of following schematic diagram. In general, bottom-up approach is meant by synthesis of nanoparticles by means of chemical reactions among the atoms/ions/molecules (**Figure 1a**). Whereas top-down involves the mechanical methods to crush/breaking of bulk into several parts to form nanoparticles (**Figure 1b**). In the next section both kind of approaches for growth of MgO nanoparticles and thin films are grown.

Properties/applications	Bulk [43, 45]	Nanoparticles [44]	Thin films [45]
Crystallite structure	Rocksalt	Rocksalt	Rocksalt
Lattice parameter (Å)	4.214	4.128	4.22
Optical band-gap (eV)	7.6	4–5	4–5

Table 1.
Properties and applications of MgO bulk, nanoparticles and thin films.

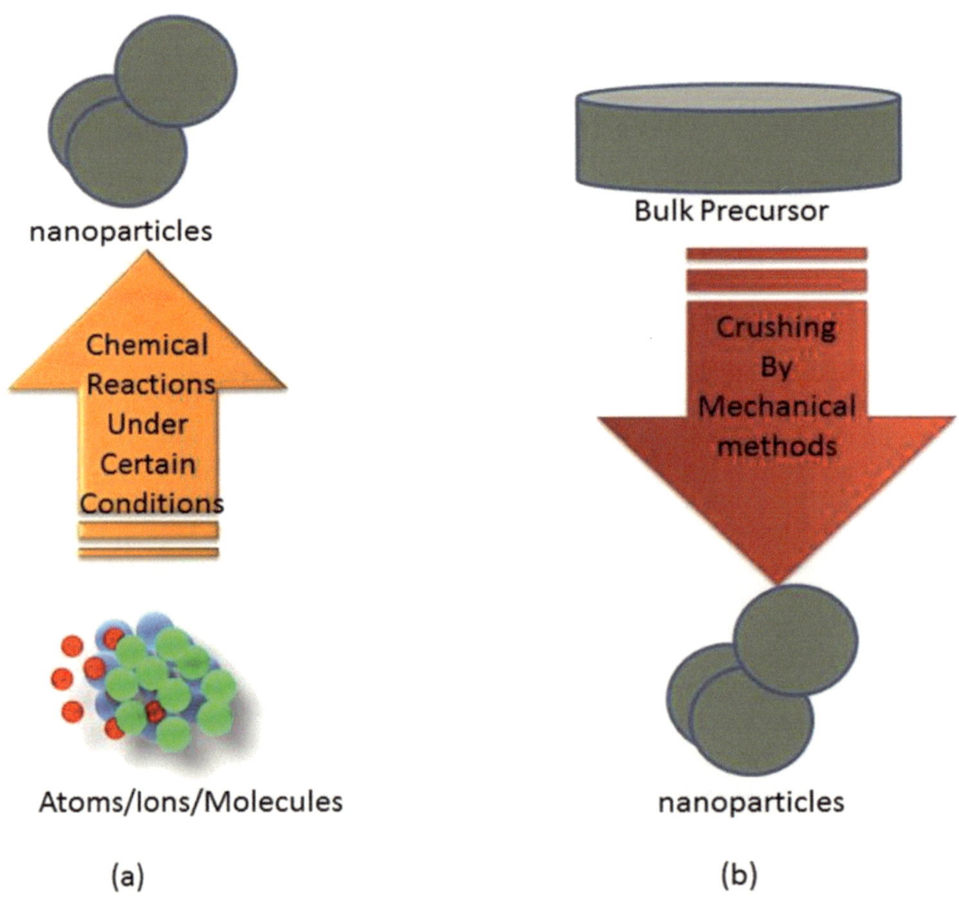

Figure 1.
Synthesis approaches for nanoparticles (a) bottom-up and (b) top-down approaches.

2. Bottom-up approaches

2.1 MgO nanoparticles

To initiate chemical reaction among the involved atoms/ions/molecules certain salts are taken as starting materials. These salts are mixed with each other to form a homogeneous solution along with a suitable chelating agent. Control of nature of solution also plays important role during synthesis process [50]. Thus, various methods are being developed by researchers to minimize annealing treatment, nature of chelating agent, pH value of solution. Some of these methods are depicted here.

Combustion synthesis is well known phenomena to synthesize nanomaterials of different kinds in its different variance [51, 52]. Most of the study utilizes solution combustion process for synthesizing nanoparticles. Typically, this method involves an oxidizer and fuel to initiate the reaction [53–55]. The most common oxidizers are metal nitrate/hydrates, ammonium nitrate and nitric acid. However, Urea, Glycine, Sucrose, Glucose, Citric Acid, Hydrazine based organic materials and Acetylacetonce are frequently used as a fuel. The water, hydrocarbons and alcohols works as solvent for reactions involved in this synthesis [53].

Thus, combustion synthesis is able to produce nanoparticle of various materials both at research purposes as well as at industrial scale [51, 53–57]. Various kind of nanoparticles like titanates [58], ferrites [59], carbonates [60], hydroxide [61] and oxides [62] are grown using this approach. Combustion synthesis is utilized for growing different kind of MgO nanostructures [63, 64] and its derivative [65–67].

Our group utilizes, this method to synthesize MgO nanoparticles using combustion synthesis while taking magnesium nitrate as an oxidizer and citric acid as fuel [68]. This method shows reproducibility [69]. The following equation is expected during synthesis process.

$$Mg\,(NO_3)_{2.4}\,H_2O + \text{Citric acid} \rightarrow MgO + CO_2 + NO + H_2O$$

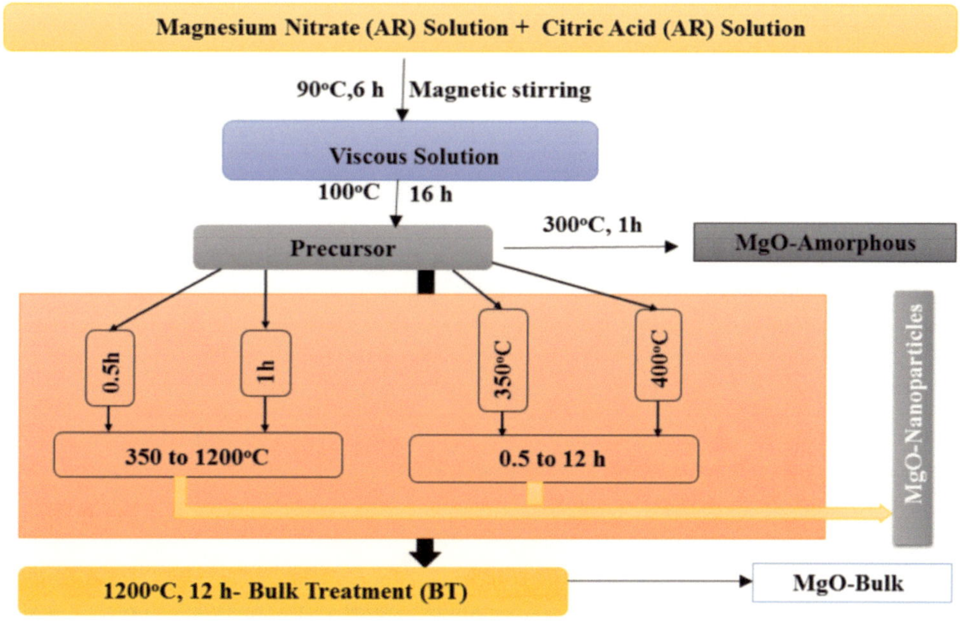

Figure 2.
Synthesis of MgO nanoparticles from magnesium nitrate Ref. [69].

Figure 2 depicts schematic diagram of synthesis process. It is clear that synthesis takes place at low temperature, which reduces cost of synthesis. **Figure 3** shows representative X-ray diffraction (XRD) pattern of the nanoparticle synthesized at 500°C for 1 h. The method is able to produce nanoparticle with pure phase and no other crystalline phases are observed [70].

Green synthesis techniques utilize natural extracts [71] as fuels/oxidizer. Some of the natural extracts for synthesizing MgO nanoparticles are Neem leaves [72], *Artemisia abrotanum* Herba Extract [73], orange fruit [74], Aqueous Eucalyptus globules leaf [75] and Medicinal Plant *Pisonia grandis* R.Br. Leaf [76].

Microwave synthesis utilizes microwave radiation rather than furnace heating in order to avoid longer duration of heating to precursor [77, 78]. This method was successfully applied to form MgO nanoparticles by number of researchers [79–81].

Other methods which are effectively used to grow nanoparticles are facile [82, 83] and miroemulsion synthesis [84–86].

2.2 MgO thin films

Spin coating method is well known tool for growing thin films which utilizes chemical reaction to form materials on the given substrate [87, 88]. Sol-gel chemistry is helpful to synthesize thin films of MgO of desired crystallographic orientation using spin coater [89, 90]. **Figure 4** shows schematic of sol-gel method utilizing a spin coater to grow thin film [91, 92].

Atomic layer deposition (ALD) method allows depositions with excellent uniformity and conformality, with a cost-effective methodology [93, 94]. Thickness and composition control are usually possible over large-area substrates. Thin films of MgO were deposited by atomic layer epitaxy (ALE) from bis(cyclopentadienyl) magnesium and water using soda lime glass and Si(100) as substrates [95]. In another study, MgO films have been grown by atomic layer deposition in the wide

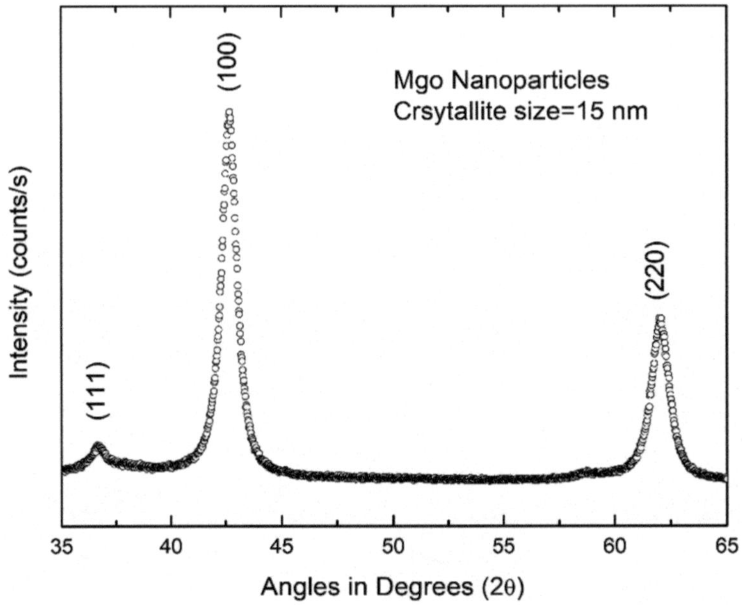

Figure 3.
X-ray diffraction pattern of the nanoparticle synthesized at 500°C for 1 h.

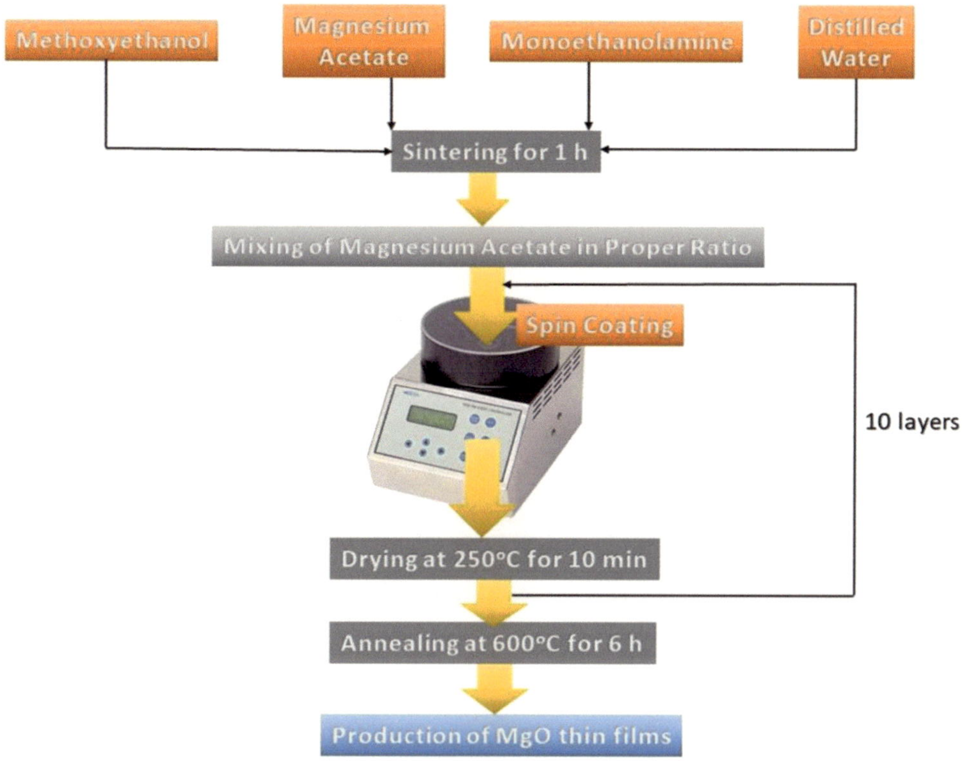

Figure 4.
Schematic of sol-gel spin coating method to grow MgO thin films. This schematic is based on the method described Ref. [91].

deposition temperature window of 80–350°C by using bis(cyclopentadienyl) magnesium and H_2O precursors [96].

3. Top-down approaches

Top-down approaches are mostly utilized to grow thin films of inorganic materials. Some of these methods are discussed here.

3.1 MgO nanoparticle

Mechanical milling/high energy ball milling is well known method which utilizes bulk counterpart as starting material and used for growing nanoparticles of different kind of materials [97, 98]. Depending upon milling process, the milling machines are categories as follows: tumbler ball mills, vibratory mills, planetary mills, and attritor mills [99, 100]. In the ball milling process, powder mixture or bulk powder placed in the ball mill is subjected to high-energy collision from the balls for nanoparticle synthesis. **Figure 5** depicts the schematic of high energy ball milling system [101]. Though this technique is effective to synthesize oxide nanoparticles [102, 103], however, no report is available for synthesizing MgO.

3.2 MgO thin films

Top-down approaches for growing MgO thin films are depicted in this section.

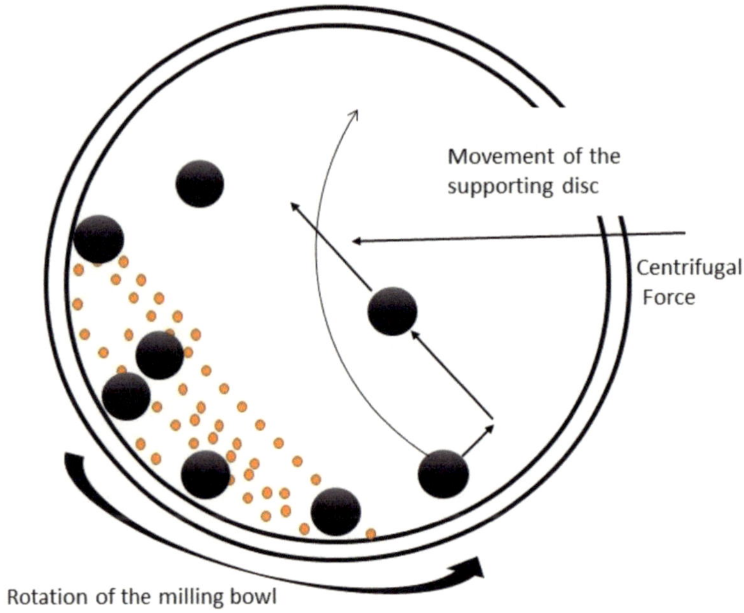

Figure 5.
Schematic of ball-milling process. Redrawn from Ref. [101].

e-Beam evaporation method involves the evaporation of material target with e-beam energy [104]. Schematic of this method is shown in **Figure 6**. This method is effectively used to grow MgO thin films on different type of substrates like NaCl [105], Si [106], fused quartz [107] as well as on metallic layers [108, 109].

Figure 7 shows the MgO thin films on fused quartz substrate along with MgO powder. Both the films of thickness around 5 and 50 nm reveal almost amorphous

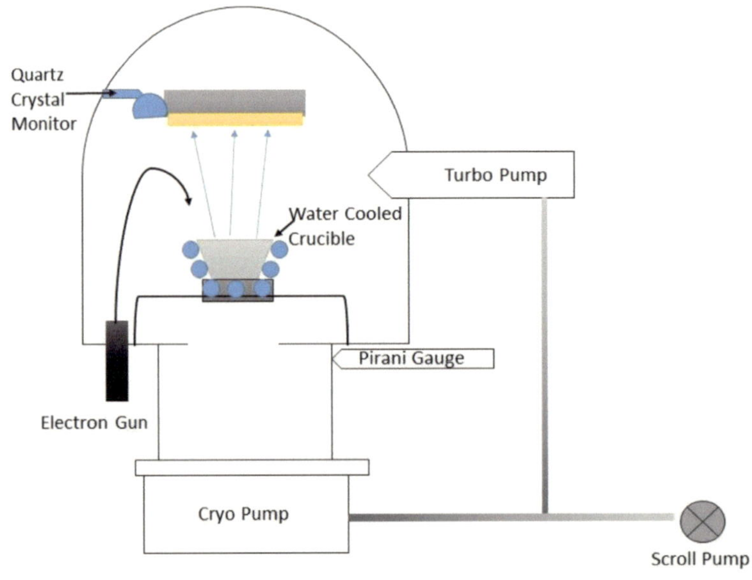

Figure 6.
e-Beam evaporation for growing MgO thin films. Schematic is based on the set-up used for growing MgO thin films in Ref. [107].

nature. Optical absorption spectra of MgO thin films exhibit onset of film formation (**Figure 8**). This method is also utilized to grow MgO thin films on Si substrate. Films grown on this substrate exhibits polycrystalline nature [110].

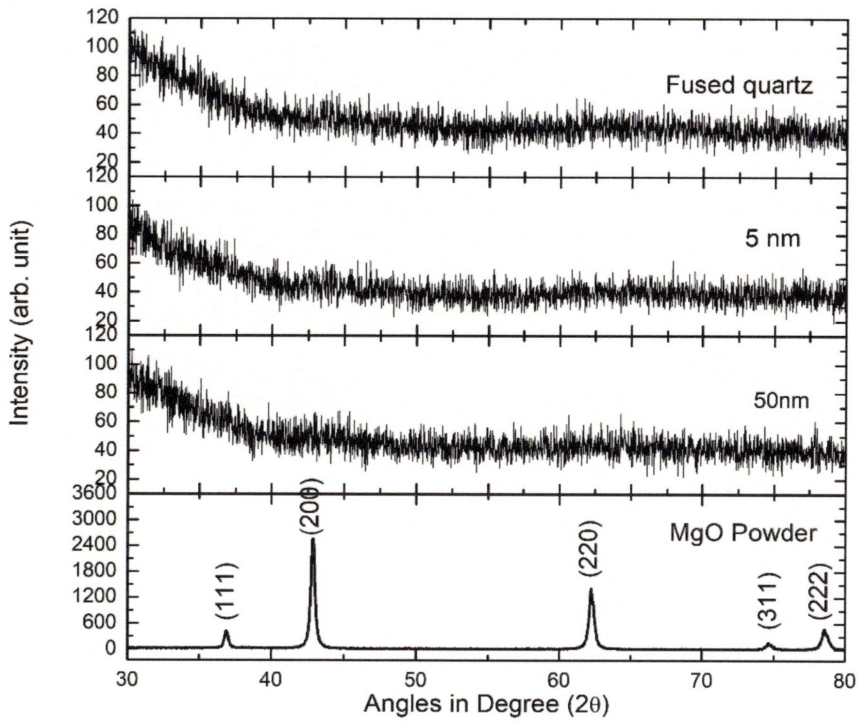

Figure 7.
X-ray diffraction pattern of MgO thin film grown on fused quartz using e-beam evaporation method.

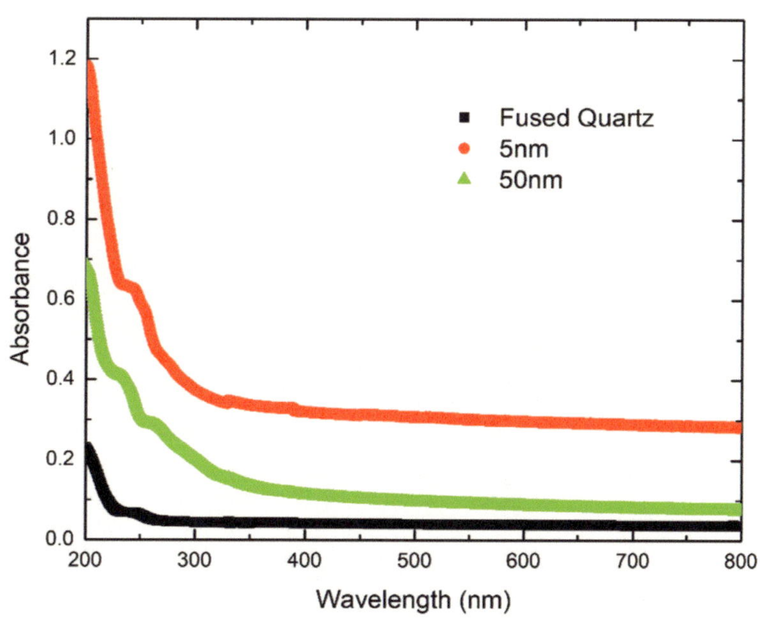

Figure 8.
UV-Vis spectra of MgO thin film grown on fused quartz using e-beam evaporation method.

Molecular beam epitaxy (MBE) utilizes e-beam for growing thin films [111]. It provides better control over stoichiometry ratio but also helpful in epitaxial growth of MgO [112, 113].

Pulsed laser deposition (PLD) as a thin film growth technique was not much popular until the late 1980s, when it has been used to grow superconducting $YBa_2Cu_3O_7$-δ films [114]. Since then, the amount of research involving this technique has increased significantly and a number of compositions have been stabilized in thin film successfully [115–122]. The schematic of PLD system is shown in **Figure 9**. In PLD, a pulsed laser beam (having wavelengths in UV range) strikes the surface of the target material to be deposited. For a short duration of laser pulse (~20 nanoseconds), enormous power (~10 MW) is delivered to the target material and absorption of energy leads to ablation before the thermodynamic equilibrium. The energy from

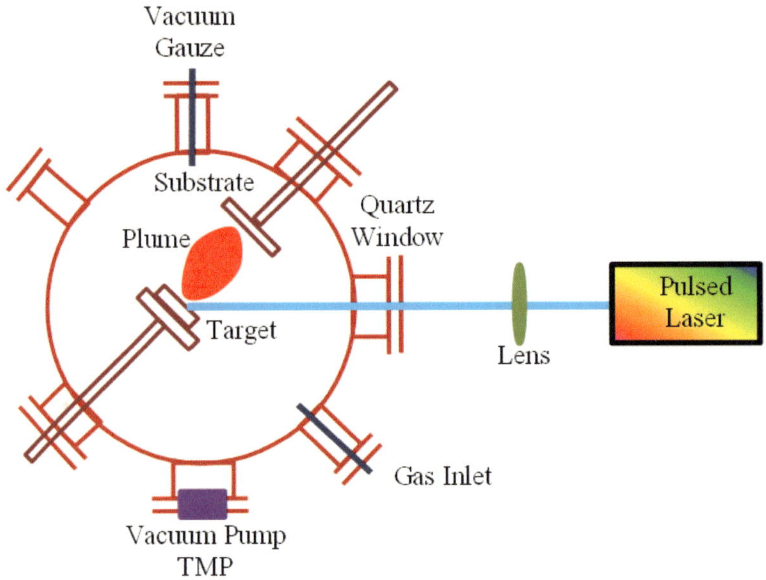

Figure 9.
Schematic of pulsed laser deposition setup.

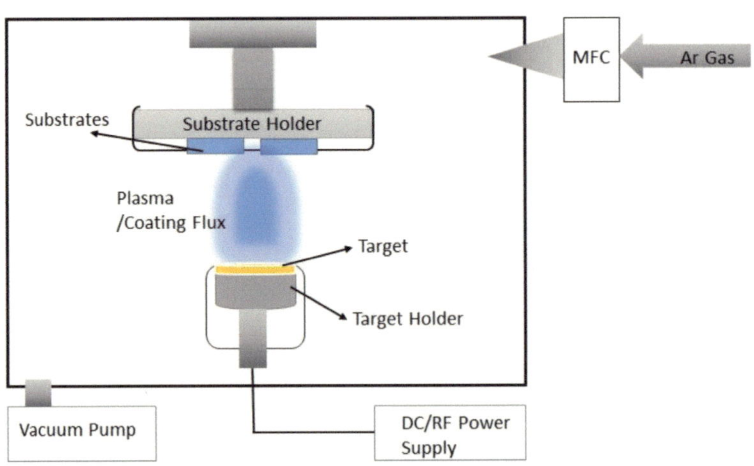

Figure 10.
Schematic of radio frequency sputtering setup based on the work reported in Ref. [131].

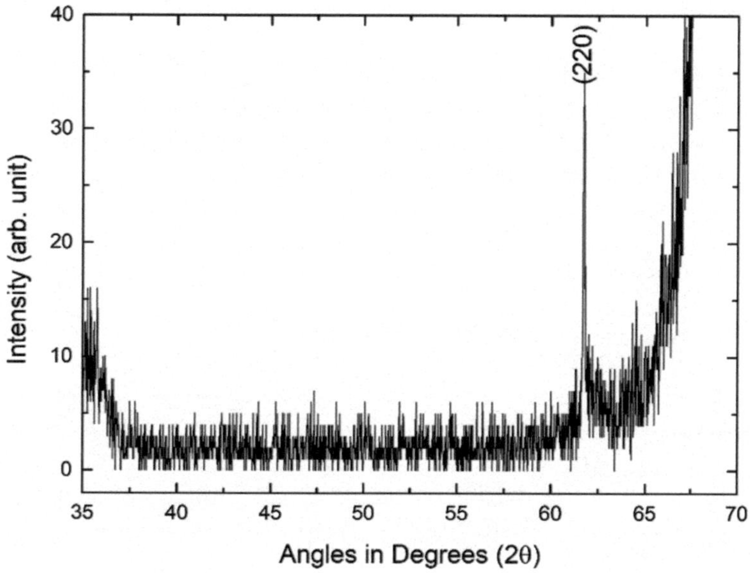

Figure 11.
XRD pattern of MgO thin film grown using rf sputtering method.

the laser evaporates the target's surface and the ablated material forms the plasma plume which finally deposit on the substrate mounted in front of the target.

The main advantage of PLD is the stoichiometric transfer from the multicomponent target to the thin film form, which otherwise is hard to achieve with any other thin film growth technique such as thermal evaporation or sputtering. The pulsed nature of PLD allows precise control on the film growth rates. Some drawbacks of the PLD technique are small area deposition, growth of macroscopic particles (particulates) during the ablation process, defects produced during growth, etc.

These advantages of this method allow researchers to grow MgO films on yttrium stabilized zirconia (111) substrates [123], Si (100) [124] and Al_2O_3 (0001) substrates [125].

Radio frequency sputtering method: At present most desired application of MgO is its utilization as a barrier for magnetic tunnel junction and rf sputtering method is preferred choice [126, 127] as well as for other applications [128]. **Figure 10** shows the rf sputtering setup for the fabrication of thin films.

MgO films on Si substrate are grown by number of researchers [129, 130] as well as by our group [42, 131]. Films grown on Si substrate are both amorphous [132, 133] and crystalline [134, 135] in nature depending upon the deposition time and annealing temperature. **Figure 11** shows the XRD pattern of the MgO thin film grown at substrate temperature of 350°C, deposition time of 400 min and annealing temperature of 800°C for 1 h followed by 300°C and 24 h.

Apart from this, number of groups utilizes **chemical deposition (CVD) method** to grow MgO thin films on different substrates [136–138].

4. Conclusions

Thus, an overview of bottom-up and top-down approaches for synthesis of MgO nanoparticles and thin films is depicted in this chapter. Chemical methods are effective to grow nanoparticles, however, later is successful to grow thin films.

Acknowledgements

JPS is helpful to Prof Ik-Jae Lee, Pohang Accelerator Laboratory, South Korea, for providing access to RF sputtering setup. GP is highly thankful to Dr. Deependra Kumar Jha, Vice Chancellor, UPES, Dehradun, for providing necessary research facilities to carry out present research work at UPES.

Conflict of interest

The authors declare no conflict of interest.

Notes/thanks/other declarations

Thanks to the editor for inviting one of authors JPS for submission of chapter. JPS and GP are thankful to the publisher to waive off the article processing charges for the chapter.

Author details

Jitendra Pal Singh[1*], Manish Kumar[1], Aditya Sharma[2*], Ganesh Pandey[3,5*], Keun Hwa Chae[4] and Sangsul Lee[1]

1 Pohang Accelerator Laboratory, Pohang University of Science and Technology, Pohang, South Korea

2 Department of Physics, Manav Rachna University, Faridabad, Haryana, India

3 University of Petroleum and Energy Studies (UPES), Dehradun, Uttarakhand, India

4 Advanced Analysis Center, Korea Institute of Science and Technology, Seoul, South Korea

5 Gus Global Services Private Limited, Gurugram, Haryana, India

*Address all correspondence to: jitendra2029@postech.ac.kr; adityaiuac@gmail.com and gp30695@gmail.com

References

[1] Salata OV. Applications of nanoparticles in biology and medicine. Journal of Nanobiotechnology. 2004;**2**:3

[2] Khan I, Saeed K, Khan I. Nanoparticles: Properties, applications and toxicities. Arabian Journal of Chemistry. 2019;**12**:908-931

[3] Mitsugi F, Nakamura A, Kodama Y, Ohkubo T, Nomoto Y. Preparation of inorganic solid state thin film for electrochromic application. Thin Solid Films. 2007;**515**:4159-4165

[4] Cauduro ALF, dos Reis R, Chen G, Schmid AK, Méthivier C, Rubahn H-G, et al. Crystalline molybdenum oxide thin-films for application as interfacial layers in optoelectronic devices. ACS Applied Materials & Interfaces. 2017;**9**:7717-7724

[5] Singh JP, Singh V, Pandey G, Sharma A, Chae KH, Lee S. Design of MgO nanostructures for diverse applications. Heliyon. 2020

[6] Sanchez C, Boissière C, Grosso D, Laberty C, Nicole L. Design, synthesis, and properties of inorganic and hybrid thin films having periodically organized nanoporosity. Chemistry of Materials. 2008;**20**:682-737

[7] Sharma D, Kanchi S, Bisetty K. Biogenic synthesis of nanoparticles: A review. Arabian Journal of Chemistry. 2019;**12**:3576-3600

[8] Nyamukamba P, Okoh O, Mungondori H, Taziwa R, Zinya S. Synthetic methods for titanium dioxide nanoparticles: A review. Titanium Dioxide - Material for a Sustainable Environment, Dongfang Yang, IntechOpen. 2018;151-175

[9] Fahmy HM, Mohamed FM, Marzouq MH, Mustafa ABE-D, Alsoudi AM, Ali OA, et al. Review of green methods of iron nanoparticles synthesis and applications. BioNanoScience. 2018;**8**:491-503

[10] Chambers SA. Epitaxial growth and properties of thin film oxides. Surface Science Reports. 2000;**39**:105-180

[11] Ramanathan, Shriram, editor. Thin Film Metal-Oxides, Fundamentals and Applications in Electronics and Energy. US: Springer; 2010

[12] Barber ZH. The control of thin film deposition and recent developments in oxide film growth. Journal of Materials Chemistry. 2006;**16**:334-344

[13] Kakade N, Kumar R, Sharma SD, Datta D. Cost effectiveness of silver nanoparticles over gold nanoparticles in nano-particle aided radiotherapy. In: Proc. of 39th Annual Conference of Association of Medical Physicists of India (AMPICON 2018); Chennai City, India

[14] Piccinno F, Gottschalk F, Seeger S, Nowack B. Industrial production quantities and uses of ten engineered nanomaterials in Europe and the world. Journal of Nanoparticle Research. 2012;**14**:1109

[15] Singh JP, Kim SH, Won SO, Lim WC, Lee I-J, Chae KH. Covalency, hybridization and valence state effects in nano-and micro-sized $ZnFe_2O_4$. CrystEngComm. 2016;**18**:2701-2711

[16] Singh JP, Kim SH, Won SO, Lee IJ, Chae KH. Atomic-scale investigation of MgO growth on fused quartz using angle-dependent NEXAFS measurements. RSC Advances. 2018;**8**:31275-31286

[17] Zhang J, Huang F, Lin Z. Progress of nanocrystalline growth kinetics based on oriented attachment. Nanoscale. 2010;**2**:18-34

[18] Yamamoto T, Adkar N, Okano Y, Ujihara T, Dost S. Numerical investigation of the transport phenomena occurring in the growth of SiC by the induction heating TSSG method. Journal of Crystal Growth. 2017;**474**:50-54

[19] Kolhatkar AG, Jamison AC, Litvinov D, Willson RC, Lee TR. Tuning the magnetic properties of nanoparticles. International Journal of Molecular Sciences. 2013;**14**:15977-16009

[20] Singh JP, Gagan D, Srivastava RC, Agrawal HM, Kumar R. Raman and Fourier-transform infrared spectroscopic study of nanosized zinc ferrite irradiated with 200 MeV Ag15+ beam. Journal of Alloys and Compounds. 2013;**551**:370-375

[21] Khodashenas B, Ghorbani HR. Synthesis of silver nanoparticles with different shapes. Arabian Journal of Chemistry. 2019;**12**:1823-1838

[22] Carney RP, Kim JY, Qian H, Jin R, Mehenni H, Stellacci F, et al. Determination of nanoparticle size distribution together with density or molecular weight by 2D analytical ultracentrifugation. Nature Communications. 2011;**2**:335

[23] Zhang H, Minnich AJ. The best nanoparticle size distribution for minimum thermal conductivity. Scientific Reports. 2015;**5**:8995

[24] Hu L, Hecht DS, Grüner G. Carbon nanotube thin films: Fabrication, properties, and applications. Chemical Reviews. 2010;**110**:5790-5844

[25] Loureiro J, Santos JR, Nogueira A, Wyczisk F, Divay L, Reparaz S, et al. Nanostructured p-type Cr/V_2O_5 thin films with boosted thermoelectric properties. Journal of Materials Chemistry A. 2014;**2**:6456-6462

[26] Ma C, Liu M, Chen C, Lin Y, Li Y, Horwitz JS, et al. The origin of local strain in highly epitaxial oxide thin films. Scientific Reports. 2013;**3**:3092-1-5

[27] Shayesteh SF, Dizgah AA. Effect of doping and annealing on the physical properties of ZnO:Mg nanoparticles. Pramana-Journal of Physics. 2013;**81**:319-330

[28] Singh JP, Srivastava RC, Agrawal HM, Kushwaha RPS, Chand P, Kumar R. EPR study of nanostructured zinc ferrite. International Journal of Nanoscience. 2008;**7**:21-27

[29] Neagu D, Tsekouras G, Miller DN, Menard H, Irvine JTS. In situ growth of nanoparticles through control of nonstoichiometry. Nature Chemistry. 2013;**5**:916-923

[30] Gunnarsson R, Helmersson U, Pilch I. Synthesis of titanium-oxide nanoparticles with size and stoichiometry control. Journal of Nanoparticle Research. 2015;**17**:353

[31] Phillips JM. Substrate selection for thin-film growth. MRS Bulletin. 1995;**20**:35-39

[32] Wu S-j J, Houng B, Huang B-s. Effect of growth and annealing temperatures on crystallization of tantalum pentoxide thin film prepared by RF magnetron sputtering method. Journal of Alloys and Compounds. 2009;**475**:488-493

[33] Dualeh A, Tétreault N, Moehl T, Gao P, Khaja Nazeeruddin M, Grätzel M. Effect of annealing temperature on film morphology of organic-inorganic hybrid pervoskite solid-state solar cells. Advanced Functional Materials. 2014;**24**:3250-3258

[34] García-Valenzuela JA, Andreu J, Bertomeu J. Effect of the base pressure

achieved prior deposition on the main properties of ZnO:Al films obtained by DC magnetron sputtering at room temperature for electrical contact use. Journal of Vacuum Science & Technology A. 2017;**35**:021603

[35] Hua Q, Ligang W, Ruijin L, Wenfeng Y. Influence of target substrate distance on the properties of transparent conductive Si doped ZnO thin films. Optik. 2014;**125**:3902-3907

[36] Wang C, Cheng BL, Wang SY, Lu HB, Zhou YL, Chen ZH, et al. Effects of oxygen pressure on lattice parameter, orientation, surface morphology and deposition rate of $(Ba_{0.02}Sr_{0.98})TiO_3$ thin films grown on MgO substrate by pulsed laser deposition. Thin Solid Films. 2005;**485**:82-89

[37] Agrawal A, Habibi HR, Agrawal RK, Cronin JP, Roberts DM, Caron-Popowich R'S, et al. Effect of deposition pressure on the microstructure and electrochromic properties of electron-beam-evaporated nickel oxide films. Thin Solid Films. 1992;**221**:239-253

[38] Pitt KEG, Howard AJ. The nature of residual gases during the deposition of resistive thin films. Vacuum. 1968;**18**:517-518

[39] Weber T P, Ma B, Balachandran U, McNallan M. Fabrication of biaxially textured magnesium oxide thin films by ion-beam-assisted deposition. Thin Solid Films. 2005;**476**:79-83

[40] Marton Z, Seo SSA, Egami T, Lee HN. Growth control of stoichiometry in $LaMnO_3$ epitaxial thin films by pulsed laser deposition. Journal of Crystal Growth. 2010;**312**:2923-2927

[41] Baeumer C, Xu C, Gunkel F, Raab N, Heinen RA, Koehl A, et al. Surface termination conversion during $SrTiO_3$ thin film growth revealed by X-ray photoelectron spectroscopy. Scientific Reports. 2015;**5**:11829

[42] Singh JP, Ji MJ, Kumar M, Lee IJ, Chae KH. Unveiling the nature of adsorbed species onto the surface of MgO thin films during prolonged annealing. Journal of Alloys and Compounds. 2018;**748**:355-362

[43] Shand MA. Physical and chemical properties of magnesium oxide. In: Shand MA, editor. The Chemistry and Technology of Magnesia. John Wiley & Sons, Ltd. 2006

[44] Singh J, Chae K. d° Ferromagnetism of magnesium oxide. Condensed Matter. 2017;**2**:36

[45] Płócienni P, Guichaoua D, Zawadzka A, Korcala A, Strzelecki J, Trzaska P, et al. Optical properties of MgO thin films grown by laser ablation technique. Optical and Quantum Electronics. 2016;**48**:277

[46] Choudhury B, Choudhury A. Microstructural, optical and magnetic properties study of nanocrystalline MgO. Materials Research Express. 2014;**1**:025026

[47] Raja AME, Jayachandran M, Sanjeeviraja C. Fabrication techniques and material properties of dielectric MgO thin films—A status review. CIRP Journal of Manufacturing Science and Technology. 2010;**2**:92-113

[48] Iqbal P, Preece JA, Mendes PM. Nanotechnology: The "top-down" and "bottom-up" approaches. In: Gale PA, Steed JW, editors. Supramolecular Chemistry. John Wiley & Sons, Ltd. 2012

[49] Wang Y, Xia Y. Bottom-up and top-down approaches to the synthesis of monodispersed spherical colloids of low melting-point metals. Nano Letters. 2004;**4**:2047-2050

[50] Sierra-Pallares J, Huddle T, García-Serna J, Alonso E, Mato F, Shvets I, et al. Understanding bottom-up continuous hydrothermal synthesis of nanoparticles using empirical measurement and computational simulation. Nano Research. 2016;**9**:3377-3387

[51] Patil KC, Aruna ST, Mimani T. Combustion synthesis: An update. Current Opinion in Solid State and Materials Science. 2002;**6**:507-512

[52] Chae S, Lee H, Pikhits PV, Kim C, Shin S, Kim DH, et al. Synthesis of terraced and spherical MgO nanoparticles using flame metal combustion. Powder Technology. 2017;**305**:132-140

[53] Varma A, Mukasyan AS, Rogachev AS, Manukyan KV. Solution combustion synthesis of nanoscale materials. Chemical Reviews. 2016;**116**:14493-14586

[54] Li F-t, Ran J, Jaronie M, Qiao SZ. Solution combustion synthesis of metal oxide nanomaterials for energy storage and conversion. Nanoscale. 2015;7:17590-17610

[55] Orante Barrón VR, Oliveira LC, Kelly JB, Milliken ED, Denis G, Jacobsohn LG, et al. Luminescence properties of MgO produced by solution combustion synthesis and doped with lanthanides and Li. Journal of Luminescence. 2011;**131**:1058-1065

[56] Aruna ST, Mukasyan AS. Combustion synthesis and nanomaterials. Current Opinion in Solid State & Materials Science. 2008;**12**:44-50

[57] Hwang CC, Tsai JS, Huang TH, Peng CH, Chen SY. Combustion synthesis of Ni-Zn ferrite powder—Influence of oxygen balance value. Journal of Solid State Chemistry. 2005;**178**:382-389

[58] Sukpanish P, Lertpanyapornchai B, Yokoi T, Ngamcharussrivichai C. Lanthanum-doped mesostructured strontium titanates synthesized via sol-gel combustion route using citric acid as complexing agent. Materials Chemistry and Physics. 2016;**181**:422-431

[59] Singh JP, Won SO, Lim WC, Lee I-J, Chae KH. Electronic structure studies of chemically synthesized $MgFe_2O_4$ nanoparticles. Journal of Molecular Structure. 2016;**1108**:444-450

[60] Singh JP, Ji M-J, Shim C-H, Kim SO, Chae KH. Effect of precursor thermal history on the formation of amorphous and crystalline calcium carbonate. Particuology. 2017;**33**:29-34

[61] Singh JP, Kim SH, Lim WC, Won SO, Chae KH. Local electronic structure investigation of the sol-gel processed calcium hydroxide material. Advanced Materials and Processes. 2018;**3**:377-381

[62] Bhardwaj R, Singh JP, Chae KH, Goyal N, Gautam S. Electronic and magnetic structure investigation of vanadium doped ZnO nanostructure. Vacuum. 2018;**158**:257-262

[63] Li S. Combustion synthesis of porous MgO and its adsorption properties. International Journal of Industrial Chemistry. 2019;**10**:89-96

[64] Nassar MY, Mohamed TY, Ahmed IS, Samir I. MgO nanostructure via a sol-gel combustion synthesis method using different fuels: An efficient nano-adsorbent for the removal of some anionic textile dyes. Journal of Molecular Liquids. 2017;**225**:730-740

[65] Sangeeta M, Karthika KV, Ravishankar R, Anantharaju KS, Nagabhushana H, Jeetendra K, et al. Synthesis of ZnO, MgO and ZnO/MgO by solution combustion method: Characterization and photocatalytic studies. Materials Today: Proceedings. 2017;**4**:11791-11798

[66] Ning P, Zhang F, Wang LJ, Zhou Y, Wang YJ, Wu YY, et al. Sol-gel derived AgMgO films for antibacterial and bioactive surface modification of niobium metal. Materials Chemistry and Physics. 2020;**243**:122646

[67] Vahid BR, Haghighi M. Urea-nitrate combustion synthesis of $MgO/MgAl_2O_4$ nanocatalyst used in biodiesel production from sunflower oil: Influence of fuel ratio on catalytic properties and performance. Energy Conversion and Management. 2016;**126**:362-372

[68] Singh JP, Won SO, Lim WC, Shim C-H, Chae KH. Optical behavior of MgO nanoparticles investigated using diffuse reflectance and near edge X-ray absorption spectroscopy. Materials Letters. 2017;**198**:34-37

[69] Singh JP, Chae KH. Local electronic structure perspectives of nanoparticle growth: The case of MgO. ACS Omega. 2019;**4**:7140-7150

[70] Singh JP, Lim WC, Won SO, Song J, Chae KH. Synthesis and characterization of some alkaline-earth-oxide nanoparticles. Journal of the Korean Physical Society. 2018;**72**:890-899

[71] Iravani S. Green synthesis of metal nanoparticles using plants. Green Chemistry. 2011;**13**:2638-2650

[72] Krishna Moorthy S, Ashok CH, Venkateswara Rao K, Viswanathana C. Synthesis and characterization of Mgo nanoparticles by neem leaves through green method. Materials Today: Proceedings. 2015;**2**:4360-4368

[73] Suresh J, Yuvakkumar R, Sundrarajan M, Hong SI. Green synthesis of magnesium oxide nanoparticles. Advanced Materials Research. 2014;**952**:141-144

[74] Ganapathi Rao K, Ashok CH, Venkateswara Rao K, Shilpa Chakra CH, Akshaykranth A. Eco-friendly synthesis of MgO nanoparticles from orange fruit waste. International Journal of Advanced Research in Physical Science. 2015;**2**:1-6

[75] Jeevanandam J, Chan YS, Ku YH. Aqueous *Eucalyptus globulus* leaf extract-mediated biosynthesis of MgO nanorods. Applied Biological Chemistry. 2018;**61**:197-208

[76] Joghee S, Ganeshan P, Vincent A. Ecofriendly biosynthesis of zinc oxide and magnesium oxide particles from medicinal plant *Pisonia grandis* R.Br. Leaf extract and their antimicrobial activity. BioNanoScience. 2019;**9**:141-154

[77] Gerbec JA, Magana D, Washington A, Strouse GF. Microwave-enhanced reaction rates for nanoparticle synthesis. Journal of the American Chemical Society. 2005;**127**:15791-15800

[78] Motshekga SC, Pillai SK, Ray SS, Rui KJ, Krause WM. Recent trends in the microwave-assisted synthesis of metal oxide nanoparticles supported on carbon nanotubes and their applications. Journal of Nanomaterials. 2012;**2012**:691503

[79] Makhluf S, Dror R, Nitzan Y, Abramovich Y, Jelinek R, Gedanken A. Microwave-assisted synthesis of nanocrystalline MgO and its use as a bacteriocide. Advanced Functional Materials. 2005;**15**:1708-1715

[80] Ribeiro DV, Paula GR, Morelli MR. Use of microwave oven in the calcination of MgO and effect on the properties of magnesium phosphate cement. Construction and Building Materials. 2019;**198**:619-628

[81] Mirzaei H, Davoodnia A. Microwave assisted sol-gel synthesis of MgO nanoparticles and their catalytic

activity in the synthesis of Hantzsch 1,4-dihydropyridines. Chinese Journal of Catalysis. 2012;**33**(9-10):1502-1507

[82] Marwaha N, Gupta BK, Verma R, Srivastava AK. Facile synthesis and characterization of pH-dependent pristine MgO nanostructures for visible light emission. Journal of Materials Science. 2017;**52**:10480-10484

[83] Chamack M, Mahjoub AR, Hosseinian A. Facile synthesis of nanosized MgO as adsorbent for removal of Congo red dye from wastewater. Nanochemistry Research. 2018;**3**:85-91

[84] He Y. MgO nanostructured microspheres synthesized by an interfacial reaction in a solid stabilized emulsion. Materials Letters. 2006;**60**:3511-3513

[85] Li S, Zhou B, Ren B, Xing L, Dong L, Li J. Preparation of MgO nanomaterials by microemulsion-based oil/water interface precipitation. Materials Letters. 2016;**171**:204-207

[86] Bumajdad A, Al-Ghareeb S, Madkour M, Al Sagheer F, Zaki MI. Synthesis of MgO nanocatalyst in water-in-oil microemulsion for CO oxidation. Reaction Kinetics, Mechanisms and Catalysis. 2017;**122**:1213-1229

[87] Kelso MV, Mahenderkar NK, Chen Q, Tubbesing JZ, Switzer JA. Spin coating epitaxial films. Science. 2019;**364**(6436):166-169

[88] Yoon J-G, Kwag YJ, Kim HK. Structural characterization of sol-gel derived MgO thin film on Si substrate. Journal of the Korean Physical Society. 1997;**31**:613-616

[89] Jung HS, Lee J-K, Kim JY, Hong KS. Synthesis of nano-sized MgO particle and thin film from diethanolamine-stabilized magnesium-methoxide. Journal of Solid State Chemistry. 2003;**175**:278-283

[90] Shin D-Y, Kim K-N. Electrical and optical properties of MgO films deposited on soda lime glass by a sol-gel process using magnesium acetate. Journal of Ceramic Processing Research. 2009;**10**:536-540

[91] Balta AK, Ertek Ö, Eker N, Okur İ. MgO and ZnO composite thin films using the spin coating method on microscope glasses. Materials Sciences and Applications. 2015;**6**:40-47

[92] Lee J, Jeong T, Yu SG, Jin S, Heo J, Yi W, et al. Secondary electron emission of MgO thin layers prepared by the spin coating method. Journal of Vacuum Science and Technology B. 2001;**19**:1366-1369

[93] George SM. Atomic layer deposition: An overview. Chemical Reviews. 2010;**110**:111-131

[94] Leskelä M, Mattinen M, Ritala M. Atomic layer deposition of opto-electronic materials. Journal of Vacuum Science & Technology B. 2019;**030801**:37

[95] Putkonen M, Sajavaarab T, Lauri NÈ. Enhanced growth rate in atomic layer epitaxy deposition of magnesium oxide thin films. Journal of Materials Chemistry. 2000;**10**:1857-1861

[96] Vangelista S, Mantovan R, Lamperti A, Tallarida G, Kutrzeba-Kotowska B, Spiga S, et al. Low-temperature atomic layer deposition of MgO thin films on Si. Journal of Physics D: Applied Physics. 2013;**46**:485304

[97] Indris S, Amade R, Heitjans P, Finger M, Haeger A, Hesse D, et al. Preparation by high-energy milling, characterization, and catalytic properties of nanocrystalline TiO_2.

The Journal of Physical Chemistry. B. 2005;**109**(49):23274-23278

[98] Salah N, Habib SS, Khan ZH, Memic A, Azam A, Alarfaj E, et al. High-energy ball milling technique for ZnO nanoparticles as antibacterial material. International Journal of Nanomedicine. 2011;**6**:863-869

[99] Yadav TP, Yadav RM, Singh DP. Mechanical milling: A top down approach for the synthesis of nanomaterials and nanocomposites. Nanoscience and Nanotechnology. 2012;**2**:22-48

[100] Suryanarayana C. Powder Metal Technologies and Applications. ASM Handbook, Vol. 7. Materials Park, OH: ASM International; 1998. pp. 80-90

[101] Synthesis of nanomaterials by high energy ball milling. Available from: http://www.understandingnano.com/nanomaterial-synthesis-ball-milling.html [Accessed: 30 December 2016]

[102] Amirkhanlou S, Ketabchi M, Parvin N. Nanocrystalline/nanoparticle ZnO synthesized by high energy ball milling process. Materials Letters. 2012;**86**:122-124

[103] Hosseini SG, Ayoman E. Synthesis of α-Fe_2O_3 nanoparticles by dry high-energy ball-milling method and investigation of their catalytic activity. Journal of Thermal Analysis and Calorimetry. 2017;**128**:915-924

[104] Singh V, Abhilash SR, Behera BR, Kabiraj D. Fabrication of thin self-supporting platinum targets using evaporation techniques. Nuclear Instruments and Methods in Physics Research A. 2011;**635**:20-23

[105] Aboelfotoh MO. Epitaxy of MgO on alkali halides with NaCI-type structure. Journal of Applied Physics. 1978;**49**:2770-2776

[106] Lee MJ, Park SY, Kim SG, Kim HJ, Moon SH, Kim JK. Effect of stress and density on the electrical and physical properties of MgO protecting layer for alternating current-plasma display panels. Journal of Vacuum Science and Technology A. 2005;**23**:1192-1196

[107] Singh JP, Sulania I, Prakash J, Gautam S, Chae KH, Kanjilal D, et al. Study of surface morphology and grain size of irradiated MgO thin films. Advanced Materials Letters. 2012;**3**:112-117

[108] Diao Z, Feng JF, Kurt H, Feng G, Coey JMD. Reduced low frequency noise in electron beam evaporated MgO magnetic tunnel junctions. Applied Physics Letters. 2010;**96**:202506-1-3

[109] Singh JP, Raju M, Asokan K, Jai P, Kabiraj D, Abhilash SR, et al. Magnetization in MgO based multilayers fabricated by e-beam evaporation. AIP Conference Proceedings. 2012;**1447**:749-750

[110] Singh JP, Chen CL, Dong CL, Prakash J, Kabiraj D, Kanjilal D, et al. Role of surface and subsurface defects in MgO thin film: XANES and magnetic investigations. Superlattices and Microstructures. 2015;**77**:313-324

[111] Yadavalli S, Yang MH, Flynn CP. Low-temperature growth of MgO by molecular-beam epitaxy. Physical Review B. 1990;**41**:7961-9963

[112] Niu F, Hoerman BH, Wessels BW. Epitaxial thin films of MgO on Si using metalorganic molecular beam epitaxy. Journal of Vacuum Science and Technology B. 2000;**18**:2146-2152

[113] Niu F, Meier AL, Wessels BW. Epitaxial growth and strain relaxation of MgO thin films on Si grown by molecular beam epitaxy. Journal of Vacuum Science and Technology B. 2006;**24**:2586-2591

[114] Chrisey DB, Hubler GH. Pulsed Laser Deposition of Thin Films. New York: Wiley Interscience; 1994

[115] Kumar M, Phase DM, Choudhary RJ, Lee HH. Structure and functionalities of manganite/cuprate thin film. Current Applied Physics. 2018;**18S**:33-36

[116] Kumar M, Choudhary RJ, Shukla DK, Phase DM. Metastable magnetic state and magnetotransport in disordered manganite thin films. Journal of Applied Physics. 2014;**115**:163904

[117] Kumar M, Choudhary RJ, Phase DM. Valence band structure of $YMnO_3$ and the spin orbit coupling. Applied Physics Letters. 2013;**102**:182902

[118] Kumar M, Choudhary RJ, Phase DM. Magnetic and electronic properties of $La_{0.7}Ca_{0.3}MnO_3$/h-$YMnO_3$ bilayer. Journal of Vacuum Science and Technology A. 2016;**34**:021506

[119] Kumar M, Choudhary RJ, Phase DM. Growth of different phases of yttrium manganese oxide thin films by pulsed laser deposition. AIP Conference Proceedings. 2012;**1447**:655

[120] Devi V, Kumar M, Choudhary RJ, Phase DM, Kumar R, Joshi BC. Band offset studies in pulse laser deposited $Zn_{1-x}Cd_xO$/ZnO hetero-junction. Journal of Applied Physics. 2015;**117**:225305

[121] Devi V, Kumar M, Kumar R, Singh A, Joshi B C, Band offset measurements in Zn1−x SbxO/ ZnO hetero-junctions. Journal of Physics D: Applied Physics. 2015;**48**:335103

[122] Panchal G, Choudhary RJ, Kumar M, Phase DM. Interfacial spin glass mediated spontaneous exchange bias effect in self-assembled $La_{0.7}Sr_{0.3}MnO_3$:NiO nanocomposite thin films. Journal of Alloys and Compounds. 2019;**796**:196-202

[123] Matsuzak K, Hosono H, Susak T. Layer-by-layer epitaxial growth of polar MgO(111) thin films. Physical Review B. 2010;**82**:033408-1-4

[124] Kaneko S, Ito T, Soga M, Motoizumi Y, Yasui M, Hirabayashi Y, et al. Growth of nanocubic MgO on silicon substrate by pulsed laser deposition. Japanese Journal of Applied Physics. 2013;**52**:01AN02

[125] Susaki T, Kumada S, Katase T, Matsuzaki K, Miyakawa M, Hosono H. Fabrication of flat MgO(111) films on Al_2O_3(0001) substrates by pulsed laser deposition. Applied Physics Express. 2009;**2**:091403-1-4

[126] Villegle JC, Radparvar PM, Yu LS, Faris SM. RF-sputter-deposited magnesium oxide films as high-quality adjustable tundel barriers. IEEE Transactions on Magnetics. 1989;**25**:1227-1230

[127] Chen X, Freitas PP. Magnetic tunnel junction based on MgO barrier prepared by natural oxidation and direct sputtering deposition. Nano-Micro Letters. 2012;**4**:25-29

[128] Lee JH, Eun JH, Park SY, Kim SG, Kim HJ. Hydration of r.f. magnetron sputtered MgO thin films for a protective layer in AC plasma display panel. Thin Solid Films. 2003;**435**:95-101

[129] Kaneko S, Funakubo H, Kadowaki T, Hirabayashi Y, Akiyama K. Cubic-on-cubic growth of a MgO(001) thin film prepared on Si(001) substrate at low ambient pressure by the sputtering method. Europhysics Letters. 2008;**81**:46001

[130] Nakano T, Fujimoto T, Baba S. Measurement of surface roughness and ion-induced secondary electron emission coefficient of MgO films prepared by high-pressure sputter deposition. Vacuum. 2004;**74**:595-599

[131] Singh JP, Kumar M, Lee IJ, Chae KH. X-ray reflectivity and near edge X-ray absorption fine structure investigations of MgO thin films. Applied Science Letters. 2017;**3**:47-52

[132] Singh JP, Lim WC, Chae KH. An interplay among the Mg^{2+} ion coordination, structural order, oxygen vacancies and magnetism of MgO thin films. Journal of Alloys and Compounds. 2019;**806**:1348-1356

[133] Singh JP, Lim WC, Lee J, Song J, Chae KH. Surface and local electronic structure modification of MgO film using Zn and Fe ion implantation. Applied Surface Science. 2018;**432**:131-139

[134] Singh JP, Lim WC, Lee I-J, Won SOK, Chae KH. Surface structure of MgO thin films revealed from X-ray reflectivity and near-edge X-ray absorption fine structure measurements. Science of Advanced Materials. 2018;**10**:1372-1376

[135] Singh JP, Kumar M, Lim WC, Lee HH, Lee YM, Lee S, et al. Growth of MgO on Si(001) by radio-frequency sputtering. Journal of Nanoscience and Nanotechnology. (In Press)

[136] Wang WB, Yang Y, Yanguas-Gil A, Noe NC, Girolami GS, Abelson JR. Highly conformal magnesium oxide thin films by low-temperature chemical vapor deposition from $Mg(H_3BNMe_2BH_3)_2$ and water. Applied Physics Letters. 2013;**102**:101605

[137] Ko JB, Kim SM. Preparation and electric characteristics of MgO films deposited by plasma-enhanced chemical vapor deposition. Journal of Ceramic Processing Research. 2009;**10**:643-646

[138] Sartori A, Habra NE, Bolzan M, Rossetto G, Sitran S, Barreca D, et al. Stability study of a magnesium β-diketonate as precursor for chemical vapor deposition of MgO. Chemistry of Materials. 2011;**23**:1113-1119

Chapter 3

Sonochemical Formation of Peracetic Acid in Batch Reactor: Process Intensification and Kinetic Study

Prashant D. Jolhe, Bharat A. Bhanvase, Satish P. Mardikar, Vilas S. Patil and Shirish H. Sonawane

Abstract

The present chapter highlights the kinetic studies for the sonochemical synthesis of peracetic acid (PAA) in a batch reactor. The effect of different operating parameters including acetic acid to hydrogen peroxide molar ratio, temperature, catalyst loading, effect of ultrasound, were studied using Amberlite IR-120H as a catalyst. The deactivation of the Amberlite IR-120H catalyst has also been studied. The experimental data were further utilized for the estimation of intrinsic reaction rate constants and equilibrium constants. From the experimental results; the optimized PAA concentration was observed for 471 mg/cm^3 catalyst loading at 40°C with acetic acid to hydrogen peroxide molar ratio equals to 1:1 in the presence of ultrasound. Results also revealed that the reaction rate was found to be significantly enhanced in the presence of ultrasound, which can be attributed to the enhanced mixing and in-situ formation of H_2O_2. The use of ultrasound drastically reduces the overall reaction time to 60 min, which is very less compared to 30 h as reported for conventional batch reactor utilizing H_2O_2 only.

Keywords: peracetic acid, ultrasound, kinetics, batch reactor, kinetics

1. Introduction

Peroxycarboxylic acids are widely employed as disinfecting/antimicrobial agents and thus hold great industrial importance [1]. Besides these properties, these are also utilized as bleaching agent for wood pulp in paper industries, for Baeyer-Villiger synthesis of fine chemicals and olefins epoxidation. Exceptional oxidative properties of peroxycarboxylic acids offer green alternative to conventional methods employing chlorine dioxide owing to nontoxicity of their decomposition products [2].

Although peracetic acid and peroxy-propionic acids are widely employed peroxycarboxylic acids, less stability and their explosive nature often possess difficulties in their synthesis. Generally, PAA can be prepared by two ways either from H_2O_2 or by oxidation of acetaldehyde which can be carried out in liquid of vapor phase [3]. The former way of liquid-phase synthesis is being used for several years due to comparative safer operations [1]. However, it suffers from slower reaction rates and thus requires involvement of catalysts. Industrial production of PAA is still being carried in the presence of sulfuric acid as acid catalyst [4, 5]. The utilization of sulfuric acid brings lot of challenges in view point of corrosion, environmental threats, and post-reaction catalyst recovery. Sustainability of industrial production is generally based on safety aspects of the production of chemicals and their environmental impact. The concept of green chemistry offers solution for prevention of waste products, suppresses the energy consumption, and provides safer processes with optimized productivity. Process intensification as one of the tool of green chemistry is any improvement made in chemical process that gives a considerable cleaner and more energy solving technology [6].

Several approaches have been adopted by many researchers for the process intensification in chemical process in industries. Recently, ultrasound has proven to be a promising tool for process intensification leading to enhanced reaction rates and mass transfer. The use of ultrasound offers various applications, which includes cleaning, organic synthesis, catalysis, extraction, emulsification, material processing, food processing, waste-water treatments, etc. [1, 7]. Ultrasonic waves cannot couple directly with molecular energy levels. The influence of ultrasound on a chemical reaction is attributed to the formation of cavitations. So, as it is based on the important phenomenon of cavitations which effects generation of extreme conditions of very high temperatures and pressures (1000 atm pressure and 10,000 K temperature) locally with overall ambient operating conditions. The formation of cavitations depends not only on ultrasonic power applied but also on physical properties of irradiated liquid rather than on chemical properties. Under ultrasound treatment, the course of chemical reaction can be influenced by two ways: mechanically due to streaming caused by a collapse of cavitation and by the formation of reactive species of volatile substances in the medium [8]. Due to this phenomenon, release of highly reactive free radicals, generation of turbulence and liquid circulation (acoustic streaming) takes place which in turn enhances the rates of transport processes [9]. These above mentioned effects of ultrasound have a several extensive applications in the areas of chemical processing that result in the decreased reaction time, increased reaction yield, and increase in the effectiveness of the catalyst [10–13].

Most common method for synthesis of peroxycarboxylic acids (PCA) is the oxidation of the parent aldehyde or carboxylic acid [14]. Generally, acid catalysts are used either in homogeneous or heterogeneous forms during the synthesis of peroxycarboxylic acids. Furthermore, several techniques for the preparation of per carboxylic acids have been reported in the literature that uses the batch process [3, 15, 16]. However, the formation of peroxycarboxylic acid is limited due to its unstable nature and reversible hydrolysis reaction. Additionally, the drawbacks of using sulfuric acid are corrosion of the equipment, setup of an energy-consuming distillation system for the catalyst separation, and threat to the environment [17]. Therefore, more research attention is being driven to the development of heterogeneous catalysts. The use of ultrasound during the preparation of peroxycarboxylic acids is an important issue of investigation.

Ultrasound irradiation has been proved to be environmentally benign method to process several chemical reactions [18, 19]. The use of ultrasound enhances the reaction rate which is supposed to be due to H_2O_2 generation and intense mixing which enhances the mass transfer rate [20]. Considering the above discussion herein present report, we are demonstrating the synthesis of PAA in presence of ultrasonic irradiations. The cation-exchange acid resin- Amberlite IR 120H was used as catalyst. The effects of several parameters such as molar ratio of acetic acid to hydrogen peroxide, temperature, catalyst loading, and ultrasonic irradiations were investigated.

DOI: http://dx.doi.org/10.5772/intechopen.89268

2. Experimental

2.1 Materials

PAA was synthesized using anhydrous acetic acid and hydrogen peroxide (30 wt. %) (S.D. Fine Chem.) in presence of cation-exchange resin as a catalyst. Cation-exchange resin (Amberlite IR-120H, mean diameter = 0.83 mm, density = 1.28 g/cm [3]) was purchased from Rohm & Haas. Analytical grade ceric sulfate, potassium iodide, and sodium thiosulfate were procured from Merck and were used as received. Sulfuric acid solution, starch indicator solution, and ferroin (1, 10-phenanthroline ferrous sulfate, 0.01 molar in water) solutions were used for the analysis of PAA. Deionized water was used in all the experiments.

2.2 Experimental setup and procedure

The experimental setup used for the synthesis is as shown in **Figure 1**. A round bottom flask of 50 mL was used for all the experiments. The reactor was kept in a commercially available 1.5 L ultrasound bath (Biotechnique Laboratory, Model BTC Sr. No. 10242, frequency = 40 kHz, power = 150 W, amplitude = 15) equipped with external cryostat for maintaining a constant temperature (20–40°C). During the course of reaction, 50 mL round bottom flask containing 13.4 mL of reaction mass was irradiated in ultrasound bath. The cation-exchange capacity of Amberlite IR-120H by dry weight (meq/g) was used to calculate the apparent $[H^+]$.

In a typical reaction, 4.5 ml (0.074 moles) of glacial acetic acid and 8.9 ml (0.078 moles) of 30% H_2O_2 solution were thoroughly mixed in a 50 mL round bottom flask. To this solution, certain amount (0.75–1.5 g) of dry catalyst, that is, Amberlite IR-120H was added. The resultant reaction mixture was then treated for 60 min using ultrasound irradiation (40 kHz, 150 W). A fixed amount of reaction mixture (0.3 mL) was sampled at regular interval of time and analyzed for the determination of PAA concentration. Mole fractions of H_2O_2 and PAA were determined by the titration method previously reported by Greenspan and Mackellar [21]. The effect of different parameters, namely effect of ultrasound, PAA: H_2O_2 mol ratios, effect of temperature, effect of catalyst concentration, and catalyst deactivation were studied. The molar ratio of acetic acid to hydrogen peroxide was varied from 1:0.45 to 1:2.50. All the experiments were performed in the temperature range of 20–40°C with an interval of 10°C and at the atmospheric pressure.

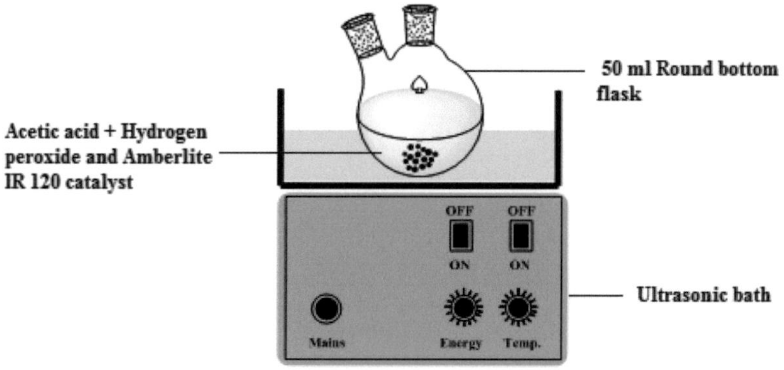

Figure 1.
Schematics of experimental setup for the preparation PAA in batch reactor.

3. Result and discussion

3.1 Effect of molar ratio of acetic acid: H_2O_2 on PAA formation

Figure 2 depicts the effect of different molar ratios of acetic acid to H_2O_2 on PAA formation. From figure, it can be observed that, as the molar ratio of acetic acid to H_2O_2 increases from 1:0.45 to 1:2.50, the PAA formation is found to be significantly increased in presence of ultrasound (40 kHz, 150 W). The energy input is improved by harnessing the power of pressurized water as transfer medium, which transfers the ultrasonic energy indirectly into the microstructured device. The mechanical and vibrational effects of the ultrasound are the main reason of better mixing as it increases the interfacial area between the phases of a heterogeneous system rather than increase in the temperature. These factors might be responsible for the increased concentration of PAA (from 2.125 to 5.2 mol/L) with an increase in the molar ratio from 1:0.45 to 1:2.50 in the presence of ultrasound (40 kHz, 150 W). The increase in the concentration of H_2O_2 increases the probability of substrate-active-site interaction and ultrasound further enhances the rate of reaction. The ultrasound increases the interfacial area between the solid and organic phases and hence the interactions also increase resulting in higher yield of PAA.

3.2 Effect of catalyst loading on PAA formation

Figure 3 depicts the effect of amount of catalyst loading on the formation of PAA in presence of ultrasound irradiation. These reactions were carried at 40°C, while maintaining the molar ratio of acetic acid to H_2O_2 as 1:1. From **Figure 3**, the observed PAA concentration for 340 mg/cm^3 catalyst loading was found to be 2.875 mol/L after 60 min. With increase in catalyst loading amount to 707 mg/cm^3, PAA concentration was found to increase to 3.375 mol/L. This can be attributed to additional acidic sites accessible at higher loading of catalyst that significantly increases the PAA concentration and equilibrium is reached earlier. The increase in the PAA yield is attributed to the synergic effect of ultrasound. Further, the concentration of PAA at 707 mg/cm^3 catalyst loading at 40°C is established to be closer to 3.45 mol/L, which is closer to the PAA concentration obtained at 471 mg/cm^3 catalyst loading at the end of 60 min. This indicates that optimum value of the catalyst loading during

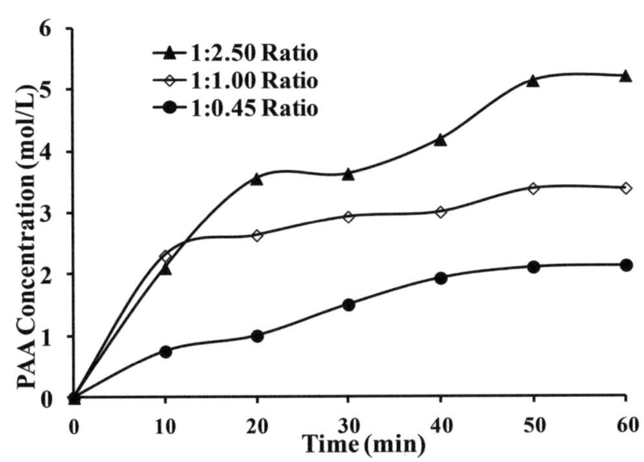

Figure 2.
Effect of molar ratio of acetic acid to hydrogen peroxide on PAA formation (catalyst loading = 471 mg/cm^3, temperature = 40°C) in batch reactor.

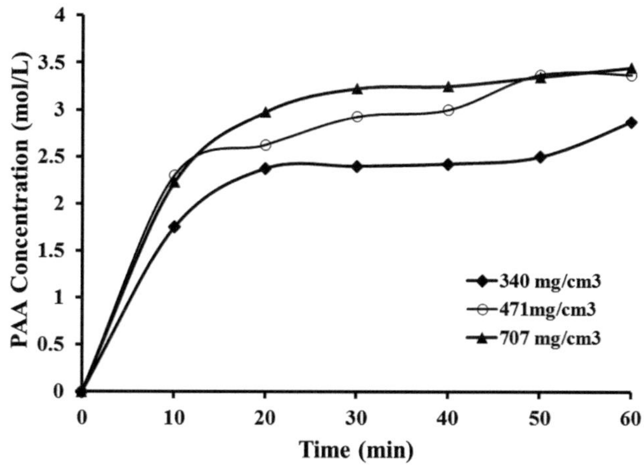

Figure 3.
Effect of Amberlite IR-120H catalyst loading (in mg/cm³) on PAA formation (molar ratio of acetic acid to H_2O_2 = 1:1, temperature = 40°C) in batch reactor.

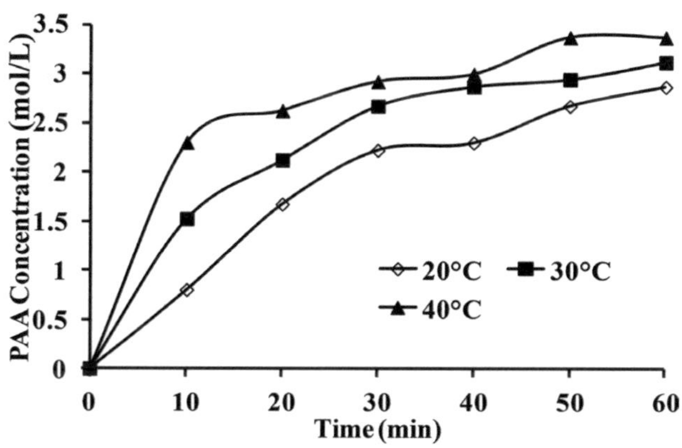

Figure 4.
Effect of reaction temperature on PAA formation (molar ratio of acetic acid to H_2O_2 = 1:1, catalyst loading = 471 mg/cm³).

the formation of PAA in batch reactor in the presence of ultrasound is 471 mg/cm^3, which provided adequate numbers of acidic condition. In case of increased catalyst loading, although, large number of acid sites is available at 707 mg/cm^3 catalyst loading, the presence of fixed amount of acetic acid limits the formation of PAA in the batch reactor. Therefore, the concentration of PAA remains unaffected. In the current set of the experiments, the equilibrium reaches approximately in 30 min against 30 h in conventional batch reactor indicating faster reaction rate and is obviously attributed to the cavitational effects of the ultrasonic irradiations, which form H_2O_2 due its chemical effect that enhances the reaction significantly.

3.3 Effect of temperature on PAA formation and kinetic study

Figure 4 depicts the effect of the temperature on PAA formation for a molar ratio of acetic acid to H_2O_2 = 1:1 and catalyst loading = 471 mg/cm^3 in batch reactor in the presence of ultrasound irradiation (40 kHz, 150 W). It is found that that the concentration of PAA increases to 3.375 mol/L (at 40°C) from 2.875 mol/L (at 20°C) at the end of 60 min. Thus, the reaction rate is enhanced with an increase in the temperature and also with the use of ultrasonic effect. The results are in good agreement with those of previous results [22]. It has been reported that the rise in the temperature provides the activation energy to the reactant molecule [23], However, too high reaction temperature may lead to decomposition of the product (PAA) leading to low product yield; thus, the experiments were carried while maintaining the temperature at 40°C maximum. Additionally, at higher temperatures, the vapor so formed may get trapped in the cavitating bubbles leading to milder collapse of cavities, which consequently result in lesser formation of H_2O_2 in the reaction medium. At lower temperature (i.e., up to 40°C), the amount of vapor present in cavitating bubble is less due to less vapor pressure. An intense collapse of cavity of the bubble brings about chemical changes, which result in the formation of more amount of H_2O_2 in the reaction medium. As a result of which, net reaction rate is enhanced at moderate reaction temperature (40°C).

Figure 5 depicts the plot of $\ln K_C$ versus 1/T for the given experimental data reported in **Figure 4**. The obtained equilibrium constant correlation for the experimental conditions is $\ln K_C = \frac{-2189.4}{T} + 9.3726$ for molar ratio of acetic acid to H_2O_2 = 1:1 and catalyst loading = 471 mg/cm^3. Also, **Figure 6** reports the plot of $\ln k_{1obs}$ and $\ln k_{2obs}$ versus 1/T for the given experimental data reported in **Figure 4**. The obtained kinetic reaction rate constant correlations are $\ln k_{1obs} = \frac{-4887.8}{T} + 7.2109$ and $\ln k_{2obs} = \frac{-2698.3}{T} - 2.1617$ for molar ratio of acetic acid to H_2O_2 = 1:1 and catalyst loading = 471 mg/cm^3. The estimated activation energy for forward reaction is 40.60 kJ/mol and for reverse reaction 22.43 kJ/mol. These activation energy values are significantly higher than previously reported values by Zhao et al. [3], where it takes 30 h to attain equilibrium in batch reactor. In present study, the equilibrium is attained within 30 min, which can be attributed to the cavitational effects of ultrasound which generates excess of H_2O_2 that accelerates the reaction rate.

3.4 Deactivation of catalyst in batch reactor

In present investigation, the deactivation and reusability of cation-exchange resin, that is, Amberlite IR-120H has also been studied. For this, batch reactions were carried in presence of H_2O_2 and ultrasound irradiation (40 kHz, 150 W). The typical experiments were performed at a molar ratio of acetic acid to H_2O_2 = 1:1 with catalyst loading of 471 mg/cm^3 and at temperature 40°C. After each

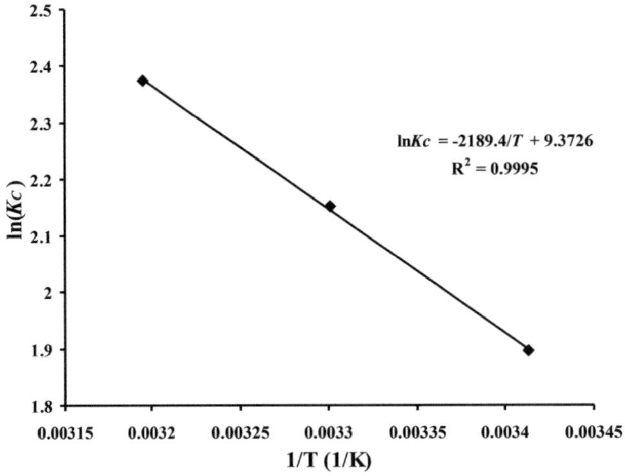

Figure 5.
The Arrhenius plot of equilibrium constant at different temperature (in kelvin) for PAA formation (molar ratio of acetic acid to H_2O_2 *= 1:1, catalyst loading = 471 mg/cm*^{3}*).*

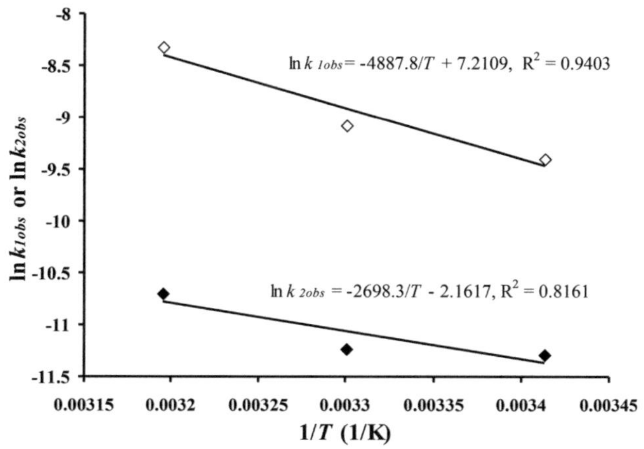

Figure 6.
The Arrhenius plots of the intrinsic reaction rate constants at different temperature (in Kelvin) for PAA formation (molar ratio of acetic acid to H_2O_2 = 1:1, catalyst loading = 471 mg/cm^3).

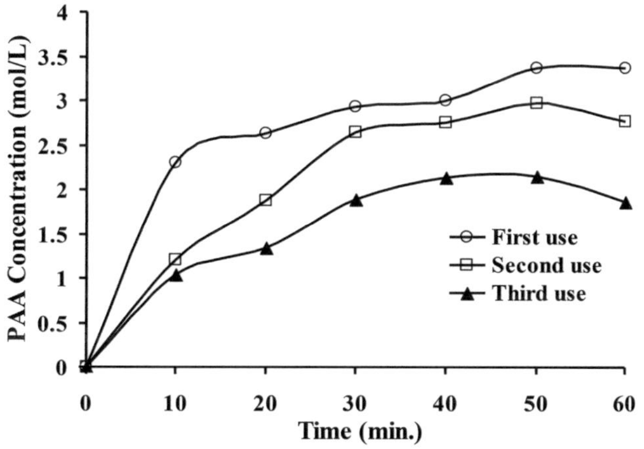

Figure 7.
Deactivation of Amberlite IR-120H catalyst during PAA formation (molar ratio of acetic acid to H_2O_2 = 1:1, catalyst loading = 471 mg/cm^3, temperature = 40°C).

experiment, the used catalyst for the synthesis of PAA was washed with water thoroughly, subsequently dried in oven at 70°C for 48 h and reused for successive experiment. **Figure 7** depicts the evaluation of the activity of fresh and reused catalyst and it has been observed that the activity of the catalyst decreases with its usage. The concentration of PAA formed in first, second, and third cycles of catalyst was observed to be 3.375, 2.975 and 2.145 mol/L, respectively, at the end of 50 min of the experimental run. The possible reasons for the deactivation of the Amberlite IR-120H catalysts are neutralization of the sulfonic acid groups, catalyst shrinkage, or loss in pore sites [23–25]. In presence of ultrasonic irradiation, as all the catalyst particles remains suspended in the reaction media, the availability of active sites of catalysts can be considered as significantly enhanced. As a result of which the greater surface area would be available for reactions leading to enhanced reaction rates. Additionally, activation of the catalyst takes place in the presence of ultrasound as ultrasound acts as cleaning agent and adsorbed material gets cleaned by ultrasound. Therefore, the activity has not been decreased drastically with the use of catalyst.

3.5 Activity of Amberlite IR-120H catalyst in batch reactor

Figure 8 depicts the comparison of the activity of cation-exchange resin, that is, Amberlite IR-120H with H_2SO_4 homogeneous catalyst and ultrasound irradiation (40 kHz, 150 W) alone (in the presence of H_2O_2) in the synthesis of peracetic acid. These experiments were carried out under same catalyst loading, that is, 471 mg/cm^3 with molar ratio of acetic acid to H_2O_2 = 1:1 at temperature 40°C. It can be seen from the **Figure 8**, the concentration of PAA is 3.375 mol/L for cation-exchange resin, that is, Amberlite IR-120H at the end of 60 min for the same condition. Further, the PAA concentration was observed to be 2.55 mol/L at the end of 60 min for same experimental conditions at 1.1% H_2SO_4 catalyst loading. Without using the catalyst, the value PAA concentration obtained is 2.225 mol/L for same experimental conditions in the presence of H_2O_2. Thus, it can be inferred that the concentration of PAA obtained is higher in case of cation-exchange resin, that is, Amberlite IR-120H in the presence of ultrasound. When only ultrasound is used, the concentration is 2.225 mol/L at the end of 60 min. It is attributed to the formation of $H^{\cdot}$ and $OH^{\cdot}$ radicals due to cavitational effects of ultrasound (Section 2.2), which accelerated the reaction rate leading to the formation of PAA.

3.6 Effect of ultrasound on PAA formation

To study the effect of ultrasound, three different sets of experiments were conducted: [1] without H_2O_2 in presence of ultrasound, [2] with H_2O_2 in presence of ultrasound, and [3] with H_2O_2 in absence of ultrasound at the same catalyst loading, that is, 471 mg/cm^3 mg with molar ratio of acetic acid to H_2O_2 = 1:1 at temperature 40°C. **Figure 9** depicts the effect of the cavitational effects of ultrasonic irradiations on the formation of H_2O_2, that is, in turn on PAA formation. The maximum PAA concentration was observed to be 0.425 mol/L in the experiment carried out without H_2O_2 in presence of ultrasound. This is recognized to the H_2O_2 formation due to the cleavage of water molecules in the presence of ultrasonic irradiations as per the mechanism provided in the Section 2.2. The rupture of water molecules results in the formation of $H^{\cdot}$ and $OH^{\cdot}$ radicals, and combination two $OH^{\cdot}$ radicals leads to the formation H_2O_2 [19], which is one of the reactant of the PAA formation reaction. This formed H_2O_2 that accelerates the PAA formation. In case of the experiments conducted in absence of ultrasound with H_2O_2, the concentration of PAA was observed to be 2.00 mol/L, which is considerably higher than that of in the presence of ultrasound without H_2O_2. Also, in the experiments conducted in presence

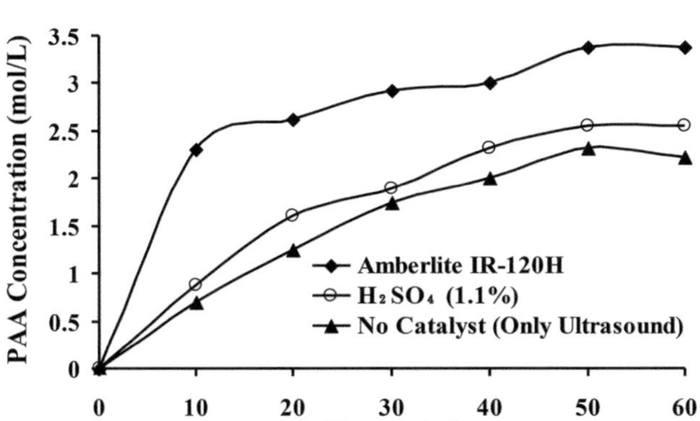

Figure 8.
Comparative study on the activity of different catalyst during PAA formation (molar ratio of acetic acid to H_2O_2 = 1:1, catalyst loading of Amberlite IR-120H = 471 mg/cm^3, temperature = 40 C).

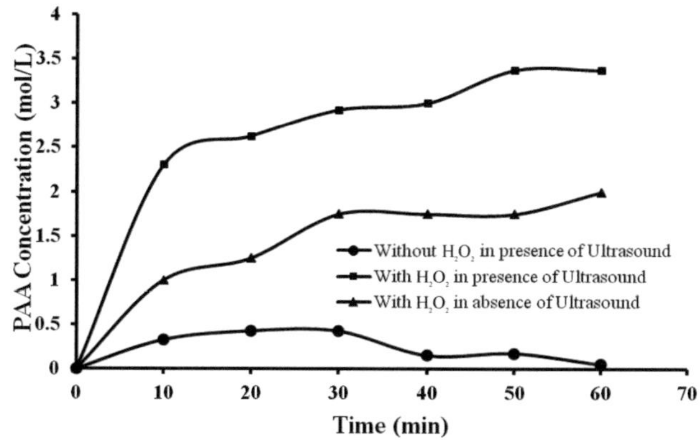

Figure 9.
PAA formation in the presence of ultrasound without H_2O_2, in presence of ultrasound with H_2O_2 and in the absence of ultrasound with H_2O_2 (molar ratio of acetic acid to H_2O_2 = 1:1, catalyst loading = 471 mg/cm^3, and temperature = 40°C).

of ultrasound with H_2O_2, the PAA concentration was observed to be significantly increased to 3.375 mol/L. This is due to the collective outcome of H_2O_2 added and generated in presence of ultrasound. Additionally, due to this surplus H_2O_2, the reaction becomes pseudo-first order and favors the PAA formation leading to suppression of reverse reaction and higher concentration of PAA in the final product.

4. Conclusions

The sonochemical synthesis of PAA in presence of Amberlite IR-120H as catalyst has been investigated. The optimum PAA concentration is obtained at temp. = 40°C, molar ratio of acetic acid to H_2O_2 1:1, and at catalyst loading of 471 mg/cm^3. Cavitational effects of ultrasonic irradiations play a significant role in the formation of PAA in batch reactor as it produces H_2O_2, which improves the reaction rate. Further, the use of batch reactor in the presence of ultrasonic irradiations enhances reaction rate significantly and reaction gets completed within merely 60 min compared to 30 h in batch reactor. The deactivation of Amberlite IR-120H catalyst occurs due to the neutralization of the sulfonic acid groups, catalyst shrinkage, or loss in pore sites. The experimental data were utilized for determination of activation energies for the synthesis and hydrolysis of PAA. The estimated activation energy for synthesis and for hydrolysis of PAA was found to be 40.6 and 22.43 kJ/mol, respectively. The activation energy for synthesis of PAA is significantly higher than the reported values in the literature. The observed effect can be attributed to the use of ultrasound during synthesis.

Author details

Prashant D. Jolhe[1], Bharat A. Bhanvase[2], Satish P. Mardikar[3], Vilas S. Patil[4] and Shirish H. Sonawane[5]*

1 Sinhgad College of Engineering, Savitribai Phule Pune University, Pune, India

2 Department of Chemical Engineering, Laxminarayan Institute of Technology, Rashtrasant Tukadoji Maharaj Nagpur University, Nagpur, India

3 Department of Chemistry, S. R. S. College, SGB Amravati University, Amravati, India

4 University Institute of Chemical Technology, North Maharashtra University, Jalgaon, India

5 Department of Chemical Engineering, National Institute of Technology, Warangal, India

*Address all correspondence to: shirishsonawne@rediffmail.com

References

[1] Sébastien Re L, Dmitry YM, Tapio S, Jyri-Pekka M, Narendra K, Kari E, et al. Synthesis of peroxypropionic acid from propionic acid and hydrogen peroxide over heterogeneous catalysts. Chemical Engineering Journal. 2009;**147**:323-329

[2] Musante RL, Grau RJ, Baltanas MA. Kinetic of liquid-phase reactions catalyzed by acidic resins: The formation of peracetic acid for vegetable oil epoxidation. Applied Catalysis A: General. 2000;**197**(11):165-173

[3] Xuebing Z, Ting Z, Yujie Z, Dehua L. Preparation of peracetic acid from hydrogen peroxide: Part I: Kinetics for peracetic acid synthesis and hydrolysis. Journal of Molecular Catalysis A: Chemical. 2007;**271**:246-252

[4] Reijo RI. Process for the preparation of peroxy acids. International Patent. WO2007031596. 2007

[5] Zhou ZX. Method for preparing peroxy acetic acid. International Patent. CN1803771. 2006

[6] Fernandez Rivas D, Angel E, Perales LW. Evaluation method for process intensification alternatives. Chemical Engineering and Processing—Process Intensification. 2018;**123**:221-232

[7] Christos V, Georgia S, Christos A. Metal Organic Frameworks (MOFs) and ultrasound: A review. Ultrasonics Sonochemistry. 2019;**52**:106-119

[8] Gogte PR, Tayal RK, Pandit AB. Cavitation: A technology on the horizon. Current Science. 2006;**91**:35-46

[9] Mason TJ, Lorimer JP. Applied Sonochemistry: The Uses of Power Ultrasound in Chemistry and Processing. Weinheim: Wiley-VCH Verlag GmbH; 2002

[10] Cains PW, Martin PD, Price CJ. The Use of Ultrasound in Industrial Chemical Synthesis and Crystallization. 1. Applications to Synthetic Chemistry. Organic Process Research and Development. 1998;**2**:34-48

[11] Balasundaram B, Harrison STL. Disruption of Brewers' yeast by hydrodynamic cavitation: Process variables and their influence on selective release. Biotechnology and Bioengineering. 2006;**94**:303-311

[12] Bhanvase BA, Pinjari DV, Sonawane SH, Gogate PR, Pandit AB. Analysis of semibatch emulsion polymerization: Role of ultrasound and initiator. Ultrasonics Sonochemistry. 2012;**19**:97-103

[13] Karekar SE, Bhanvase BA, Sonawane SH, Deosarkar MP, Pinjari DV, Pandit AB. Synthesis of zinc molybdate and zinc phosphomolybdate nanopigments by an ultrasound assisted route: Advantage over conventional method. Chemical Engineering and Processing. 2015;**87**:51-59

[14] Phillips B, Starcher PS, Ash BD. Preparation of aliphatic peroxyacids. The Journal of Organic Chemistry. 1958;**23**:1823-1826

[15] Dul'neva LV, Moskvin AV. Kinetics of formation of peroxyacetic acid. Journal of General Chemistry. 2005;**75**:1125-1130

[16] Leveneur S, Salmi T, Yu Murzin D, Estel L, Warnå J, Musakka N. Kinetic study and modeling of peroxypropionic acid synthesis from propionic acid and hydrogen peroxide using homogeneous catalysts. Industrial and Engineering Chemistry Research. 2008;**47**:656-664

[17] Leveneur S, Warnå J, Salmi T, Murzin DY, Estel L. Interaction

of intrinsic kinetics and internal mass transfer in porous ion-exchange catalysts: Green synthesis of peroxycarboxylic acids. Chemical Engineering Science. 2009;**64**:4101-4114

[18] Bhanvase BA, Pinjari DV, Gogate PR, Sonawane SH, Pandit AB. Synthesis of exfoliated poly(styrene-co-methyl methacrylate)/montmorillonite nanocomposite using ultrasound assisted in-situ emulsion copolymerization. Chemical Engineering Journal. 2012;**181-182**:770-778

[19] Bhanvase BA, Sonawane SH, Pinjari DV, Gogate PR, Pandit AB. Kinetic studies of semibatch emulsion copolymerization of methyl methacrylate and styrene in the presence of high intensity ultrasound and initiator. Chemical Engineering and Processing. 2014;**85**:168-177

[20] Gonzalez-Garcia J, Banks CE, Sljukic B, Compton RG. Electro-synthesis of hydrogen peroxide via the reduction of oxygen assisted by power ultrasound Ultrason. Sonochemistry. 2007;**14**:405-412

[21] Greenspan FP, Mackellar DG. Analysis of aliphatic per acids. Analytical Chemistry. 1948;**20**:1061-1063

[22] Varma RS, Naicket KP, Kumar DJ. Can ultrasound substitute for a phase-transfer catalyst? Triphase catalysis and sonochemical acceleration in nucleophilic substitution of alkyl halides and α-tosyloxyketones: Synthesis of alkyl azides and α-azidoketones. Journal of Molecular Catalysis A: Chemical. 1999;**149**:153-157

[23] Rajendran V, Harikumar K. Ultrasound Assisted Synthesis of Diethyl-2,2'-Thiodiacetate with 2-Bromoethylacetate Under a New Polymer-Supported Phase-Transfer Catalyst in Solid-Liquid Condition. Chemical Sciences Journal. 2015;**6**:2

[24] Russbueldt BME, Hoelderich WF. New sulfonic acid ion-exchange resins for the preesterification of different oils and fats with high content of free fatty acids. Applied Catalysis A: General. 2009;**362**:47-57

[25] Jolhe PD, Bhanvase BA, Patil VS, Sonawane SH. Sonochemical synthesis of peracetic acid in a continuous flow micro-structured reactor. Chemical Engineering Journal. 2015;**276**:91-96

Chapter 4

Ultrasonic-Assisted Cathodic Plasma Electrolysis Approach for Producing of Graphene Nanosheets

Nguyen Van Truong, Nguyen Quoc Dung, Nguyen Nhat Huy, Pham Van Hao and Dang Van Thanh

Abstract

In this chapter, we review on the production of graphene by ultrasonic-assisted cathodic plasma electrolysis approach which involves a combination process of conventional electrolysis and plasma at ambient pressure and moderate temperature. Firstly, we review on the techniques for electrochemical preparation of graphene. Then, we briefly describe plasma electrolysis approach for producing of graphene. The mechanism, advantages, and disadvantages of this technique are discussed in detail.

Keywords: ultrasonic assisted, graphene nanosheets, cathodic electrochemical discharge, plasma, electrolysis

1. Introduction

Over the past decades, graphene is one of the hottest topics in many research fields. Up to date, a numerous of possible technological applications using graphene has been explored for energy, storage, optoelectronics, energy conversion, solar cell, flexible devices, and photonics [1–3]. Graphene is currently prepared by two typical top-down and bottom-up approaches with their advantages and drawbacks. For instance, chemical vapor deposition, classified as a chemical method, is an outstanding method because graphene can be obtained with high quality. However, it needs specific substrates, high temperature and vacuum, and limited transfer requirement while producing fewer quantities due to the slow process [4]. In terms of top-down technique, the most commonly used Hummer's method for graphite chemical exfoliation requires concentrated acids and strong oxidants for pretreating and is then followed by a sonication step [5]. Unfortunately, this method has a key bottleneck such as graphene structure containing many defects or functional groups, being time-consuming, and harsh oxidation condition. Furthermore, electrochemical exfoliation is emergently proven to be promising as a sustainable and green method for large-scale graphene production since it provides a single step and is user-friendly at moderate conditions [6].

Sonoelectrochemistry, an introduction of ultrasonic irradiation into electrochemical method, has become a potential approach for synthesizing nanomaterials owing to its simple step and being environmentally friendly [7]. Fundamentally, the generation of cavitation bubbles is initiated by the propagation of pressure waves in a fluid [8, 9], and the collapse of cavitation bubbles is rapidly happening (<μs) to create local "hot spots" with high local pressure (e.g., 10,000 atm) and temperature (e.g., up to 10,000 K) [10]. During this process, highly reactive radicals of H and OH· are engendered via homolysis of water as well as other surfactants or excited state species [11]. Due to the fact that bulk graphite involves several graphene sheets which are weakly bonded through van der Waals force, there is a great possibility for taking the advantage of a tremendous energy to break this force in graphitic structure. Although the electrochemical exfoliation [12, 13] and ultrasound-assisted exfoliation [14, 15] are extensively employed for preparation of graphene and its derivative, the review of ultrasound energy and cathodic electrochemical exfoliation process for high-yield graphene production has not been explored yet.

In this part, the experimental conditions and contemporary information for producing of graphene material are only considered, with a focus on combination of electrochemical exfoliation and sonochemistry approaches. In the following parts, we firstly present an overview of electrochemical exfoliation method, and the new technique consisting of cathodic exfoliation and sonochemistry will then be comprehensively discussed.

2. Electrochemical exfoliation

Electrochemical exfoliation is one of the top-down approaches using graphite electrode as graphene precursor, which can be categorized into anodic and cathodic methods based on the electric sign of the graphite electrode. **Figure 1** displays the schematics of cathodic and anodic exfoliation mechanism. In both methods,

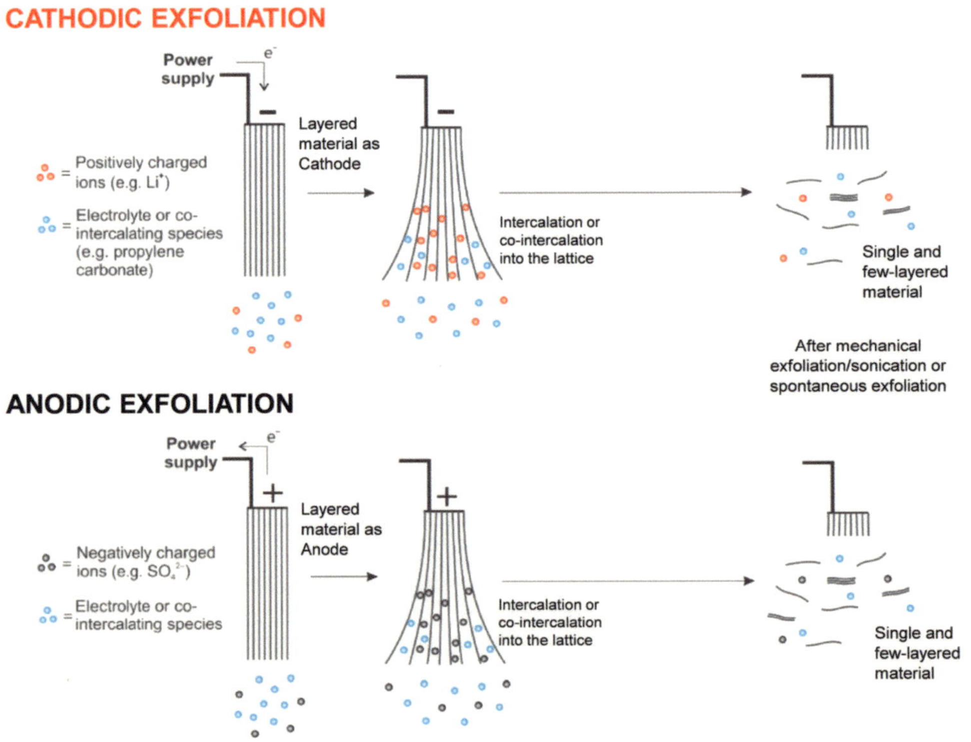

Figure 1.
Mechanism for exfoliation of graphite by cathodic and anodic methods [16].

the oppositely charged intercalating ions and the co-intercalating molecules were attracted when a charging process was generated at the working electrode [16].

2.1 Anodic exfoliation

In this method, layered carbon material is applied with a positive voltage, which drives the intercalation of negative ions existing in the electrolytic solution. Different solvents and electrolytes were employed for the exfoliation of graphene:

2.1.1 Acidic electrolytes

Using highly oriented pyrolytic graphite (HOPG) electrode, different protonic acids such as HCl, HBr, and H_2SO_4 were used for electrochemical exfoliation of graphene [6]. Among the mentioned acids, H_2SO_4 has proven to be an appropriate electrolyte due to the highest efficiency; however, the graphene nanosheets were produced with high defects/orders in their structure. In order to reduce this effect, the electrolyte was prepared by adding of 98% H_2SO_4 solution and 20% KOH solution into 10 ml water. The system was firstly applied with low voltage of +2.5 V for 1 min, and higher potential of switching between +10 and −10 V was then applied until sufficient exfoliated material is obtained. During the process, the voltage of +10 V activates the exfoliation and oxidizes the graphene flakes. The oxidized graphene is then reduced when the voltage changed back to −10 V. **Figure 2** schematically illustrates the experimental setup, and the source of graphene acts as the anode for electrochemical exfoliation.

In another work, H_2SO_4, H_3PO_4, and $H_2C_2O_4$ were used as acidic electrolytes to synthesize graphene flakes using multiple electrochemical exfoliation (MEE) methods [13]. The process of synthesis graphene flakes from spent graphite rod is

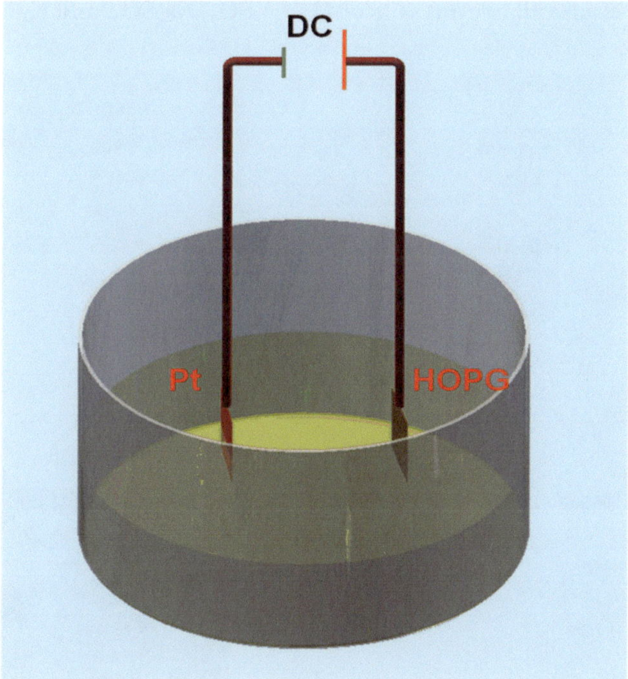

Figure 2.
Experimental setup for exfoliation of graphite by electrochemical method.

shown in **Figure 3**. In the electrochemical cell, the cathode platinum wire and the anode graphite rod were put at the top and the bottom, respectively. Upon applying a constant current at 0.1 A, the color of the solution changes gradually to black during the reaction time.

Several mild acids (i.e., phosphoric, sulfuric, oxalic, acetic, and formic acids) were employed as electrolytes for exfoliation of graphite [13]. Meanwhile, no observable exfoliation was detected in formic and acetic acids, possibly due to its small anions (e.g., HCOO– and H_3CCOO–), which results in delicate anion protection and weak expansion of graphite flakes due to ineffective intercalation of anion. Oxalic acid electrolyte is finally considered as the best choice because of its fast exfoliation and highly dispersed graphene in DI water. Furthermore, the product was easily purified by heating the exfoliated material in air.

2.1.2 Base electrolytes

There has been rare publication in electrochemical exfoliation of graphite in alkaline medium. A facile, environmentally friendly, and highly efficient exfoliation process was introduced by graphite electrolysis in a weak alkaline solution aqueous ammonium hydroxide (28–30 wt%) instead of acidic electrolyte or strong alkaline solution [17]. The graphite electrode was firstly subjected to an anodization treatment at 10 V for 30 min and subsequently treated by anodization at 10 or 5 V for 90 min. The exfoliation can be explained by the intercalation of nitrogen gas originating from the electrooxidation of ammonium hydroxide when electrolysis is performed in the alkaline electrolyte.

2.1.3 Salt electrolytes

The formation of graphite intercalation compounds (GIC) was firstly done with sulfuric acid [6]; however, sulfate ions were then realized to play a decisive role as an intercalant because all neutral salts of $(NH_4)_2SO_4$, Na_2SO_4, and K_2SO_4 exhibited

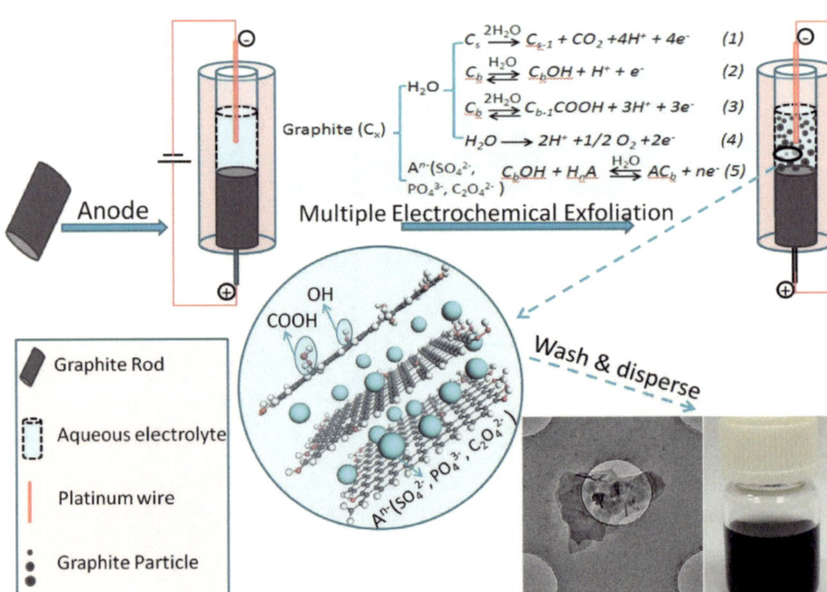

Figure 3.
Experimental setup and mechanism for production of graphene by multiple electrochemical exfoliations [13].

pronounced exfoliation efficiency [12]. In addition, the superior activity of sulfate ions can be explained by its lower reduction potential to produce SO_2 (i.e., +0.20 V). In contrast, ClO_4^- and NO_3^- ions have high reduction potentials of 1.42 and 0.96 V, respectively, to generate Cl_2 and NO gases [12]. Therefore, it is suggested that graphite sheets are exerted with large and sufficient forces to isolate weakly bonded graphite layers from each other by SO_2 and O_2 gases generated in sulfate ion oxidation.

2.2 Surfactants

Due to very excellent intercalant role of sulfate anion, the organic derivatives of sulfate ion-sulfonates such as poly(sodium-4-styrenesulfonate), sodium dodecyl sulfate (SDS), and sodium dodecyl benzene sulfonate (SDBS) were also investigated for graphite exfoliation [18–20], which play a dual role of a surfactant as well as an intercalant. The additional advantage of using surfactants in the intercalation and exfoliation process is the stable graphene suspension production because surfactants can adsorb on the surface of produced graphene layers and prevent its reassembling [20]. **Figure 4** describes how SDS with a concentration of 0.1M works

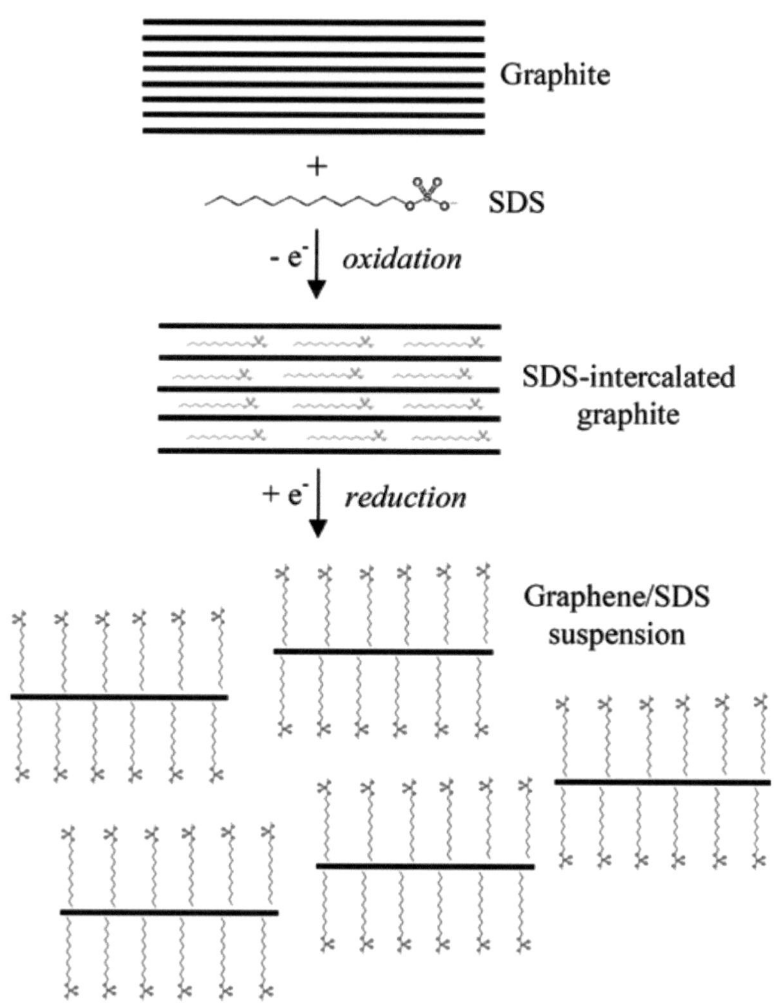

Figure 4.
Electrochemical method for graphene/SDS suspension production [20].

during graphene intercalation from the anode of graphite and exfoliation from cathode of SDS-intercalated graphite [20].

2.3 Cathodic exfoliation

The cathodic exfoliation process for graphene production has been introduced by several groups [21–24]. In this approach, the positive ions in electrolytes are intercalated into graphite interlayers by applying negative potential, which consequently facilitates their exfoliation. This process is accomplished by using either organic or aqueous solvents.

Wang et al. [21] used cathodic electrochemical exfoliation in liquid rechargeable lithium-ion batteries to prepare few-layer graphene flakes. They indicated that graphite interlayers can be intercalated by Li ions during the charging process. Subsequently, the complex was exfoliated to few-layer graphene flakes by sonication as shown in **Figure 5A**. They used a potential of −15 ± 5 V as cathodic charge in order to trigger off Li+ intercalation in graphite. The exfoliated graphene sheets had very low defects as confirmed by a small I_D/I_G ratio of 0.1 with of two to three layers of thickness and an average lateral size of 1–2 μm. Additionally, graphene nanosheets were directly obtained from electrochemical cathodic exfoliation under high negative voltages of 5–30 V using tetra-n-butylammonium (TBA) as organic electrolyte [22]. Yang et al. [22] have demonstrated that at the initial stage, the solvated TBA+ under highly negative voltage contributes to expand the layer spacing induced by the intercalation. The exfoliated graphene nanosheets exhibited high quality with the basal size of ~10 μm, three to six layers of thickness, and 0.34 nm of lattice spacing. In another example, a continuous of few-layer graphene production was developed by Abdelkader group [23] with mechanism displayed in **Figure 5B**. In their report, organic solvents and the electrolyte of lithium and alkylammonium ions (triethylammonium, Et_3NH^+) in dimethyl sulfoxide (DMSO) were used. Interestingly, they found that not only lithium ions but also Et_3NH^+ ions can be intercalated to enhance exfoliation efficiency. Furthermore, this cathodic exfoliation of graphite can avoid oxidation of graphite and decrease the defects, where few-layer graphene with no any oxidation and lateral size of 1–20 μm was produced without any sonication or centrifugation step, by using other kinds of intercalation ion such as $AlCl_4^-$ and $Al_2Cl_7^-$.

Lei et al. [24] produced graphene from graphite cathode in $AlCl_3$/EMImCl (1-ethyl-3-methylimidazolium chloride) ionic liquids using Al-ion battery. With voltage of >2.0 V, few-layer graphene can be achieved by cathodic electrolysis. The estimated mechanism for cathodic electrolytic exfoliation of graphite in ionic liquid is displayed in **Figure 5C**.

With a new term of direct cathodic exfoliation of graphite by plasma electrolysis, Thanh and his collaborators introduced a facile and rapid method to produce both graphite oxide and graphene nanosheets [25, 26]. **Figure 6A** shows the schematic of the experiment for producing plasma-electrochemically exfoliated graphene (PEEG) sheets, where high-purity graphite (HG) is employed for both the cathode and anode. The cathode tip is set above the aqueous electrolyte, while the anode is immersed in the electrolyte of KOH and $(NH_4)_2SO_4$ solution. A voltage of 60 V is applied to the system by using a DC power when the tip of cathode is immersed about 1 mm in the electrolytic solution. It is noticed that the cathode has much smaller surface area contacting to the electrolyte than the anode. As a result, a high electric field is generated at the submerged cathode tip and produces hydrogen gas bubbles via hydrolysis of water. A plasma phenomenon around the cathode tip is formed by the instant ionization of hydrogen caused by high electric field surrounding the cathode tip.

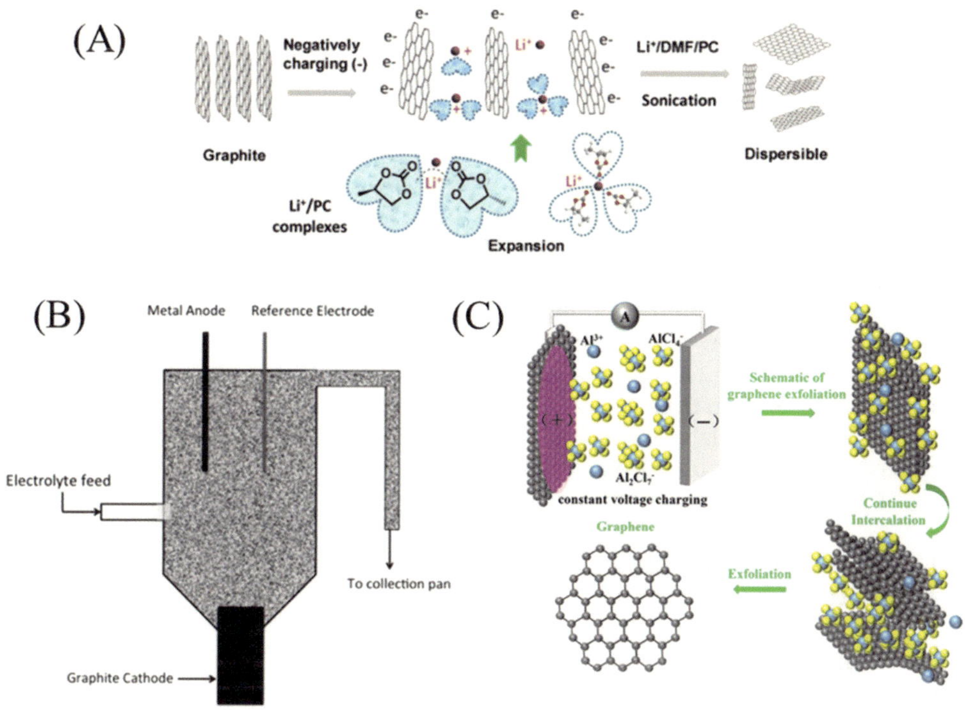

Figure 5.
Schematics of (A) electrochemical cell for a continuous process [23], (B) graphite exfoliation in the Al-ion battery [24], and (C) graphite exfoliation via intercalation [21].

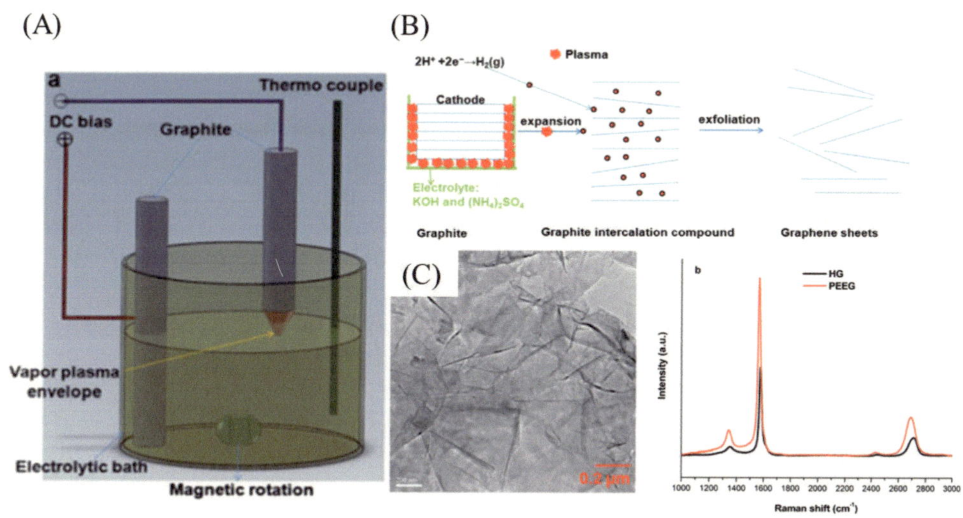

Figure 6.
(A) Experimental setup for plasma-electrochemically exfoliated graphene sheet production, (B) proposed PEEG formation mechanism, and (C) TEM image and Raman spectra of PEEG [26].

The mechanism for exfoliation of graphene sheets from graphite cathodic rod is displayed in **Figure 6B**, which attributed that the exfoliation process is cause by hydrolyzed hydrogen bubbles produced at the cathode. The explosion of hydrogen around the cathode tip expands its surface edge, which enables the insertion of hydrogen molecules into the interlayer of the graphite sheets to form intercalated graphite compounds. Simultaneously, the high temperature (e.g. ~2000°C) in instant time (e.g. nanosecond) on the cathode tip during the discharge process gives thermo-mechanical stresses on the graphite surface, which produces graphene nanosheets from graphite rod. On the other hand, graphene sheets can also be yielded from graphite rod by the electrochemical reaction occurring at the anode in a basic electrolyte medium. As a comparison, plasma-assisted electrochemical exfoliation process can produce graphene sheets with high efficiency of ~6 times faster than the conventional electrochemical method at applied voltage of 10 V. The produced material is graphene sheets with high quality as presented in the TEM image of **Figure 6C**. The structure transformation from graphite to graphene sheets can be confirmed by Raman spectra (**Figure 6C**), where the weak D band peak located at 1353 cm^{-1} is more pronounced than that in the spectrum of HG. Moreover, the 2D band peak of PEEG at 2706 cm^{-1} is shifted to lower frequency, while the intensity of PEEG is higher than that of HG. This confirms the produced graphene structure with ~2.5 μm in sheet lateral size and 2.5 nm in thickness. Interestingly, they also produced graphite oxide by this method in 2013 [25].

3. Ultrasonic-assisted cathodic electrochemical discharge exfoliation (UCEDE)

Cathodic plasma electrolysis, a combination of conventional electrolysis and plasma processes under ambient condition, employs a voltage that is much higher than that of traditional electrochemical method between two electrodes in an aqueous solution. Surface of one electrode, active electrode (workpiece, regardless of its role as anode or cathode), must be much smaller than that of the other. The method can be separated into anodic and cathodic plasma electrolysis (CPE) based on the opposition of applied voltage to the workpiece. Most of the researches have focused on the anodic regime of plasma electrolysis while little on the cathodic one [27]. CPE is principally based on the reaction or vaporization of electrolyte and the electrical breaking of gaseous envelope, resulting in the formation of sparks around the active electrode [27–29]. The comprehensive reviews of plasma electrolysis can be found in some excellent reviews [15, 25, 26].

The cathodic process has been applied for producing nitride, carbon, and other metals such as zinc, zinc-aluminum, molybdenum, and titanium-based coatings on the metal substrate [27–31]. The nanocrystalline graphite films on titanium substrate have been deposited by the cathodic plasma electrolysis from a predominant ethanol liquid phase [32, 33]. In fact, each layer of graphene in the bulk graphite is bound by weak van der Waals interactions to other two adjacent layers of hexagonally close-packed C atoms. Upon the impact on the surface as thermal extension or ultrasonication, it is readily exfoliated into separated graphene nanosheets. As a result, the plasma electrolysis phenomenon for production of graphene flakes and its derivatives is believed to be caused by the breaking of graphene layers from its bulk structure with weak van der Waals force.

Typically, using a mechanical mean such as ultrasonic after electrochemical intercalation and expansion of graphite is essential. During ultrasonication with pressure oscillations, the cavitation and shear forces, as well as the collapse of the bubbles or voids on liquid, could activate both graphite intercalation and expansion, which leads to a complete exfoliation. A new ultrasonic-assisted cathodic electrochemical discharge approach has been developed to exfoliate graphene nanosheets from graphite rod by using a combination of ultrasonic energy and in situ plasma-induced electrochemical exfoliation [15]. Without expensive ionic liquid or acidic media, this method revealed broad benefits such as direct production of graphene at ambient pressure and low temperature as well as a facile, environmentally friendly, and fast process. **Figure 7** displays the experimental setup and an actual figure for production of ultrasonic-assisted in situ plasma-induced electrochemical exfoliation graphene (UPEEG).

The formation of plasma at high voltage of 60 V is initiated by the reduction to from hydrogen gas at cathode (Eq. 1) and the oxidation to form oxygen at anode (Eq. 2):

$$2H^{+} + 2e \rightarrow H_2 \quad (1)$$

$$2H_2O \rightarrow O_2 + 4H^{+} + 4e \quad (2)$$

At appropriate voltage, the electrolyte near graphite cathode is vaporized because of the high electric field created by sharp point of cathode. This releases vigorous hydrogen around the graphite cathode tip and forms plasma zone between the cathode tip and the electrolytic solution.

Results from XPS and Raman analyses (**Figure 8A–C**) show the successful production of graphene via this method. It is believed that the positive ions surrounding the interface of electrolyte and HG tip could be speeded up directionally toward the cathode surface to implode on expanded position and to instantly exfoliate graphite at this point for graphene formation. In addition, the proposed mechanism for UPEEG formation from graphite is related to the expansion of the graphite tip in the plasma zone, where the graphene sheets were formed under the ultrasonic-assisted condition (**Figure 8D**). The extremely high-temperature outermost cathodic tip in the plasma zone leads to the expanding and cracking of graphite surrounding with weakening van der Waals forces. In this case, the violent fluctuation in the vapor plasma envelope and the cavitation from the sonication are

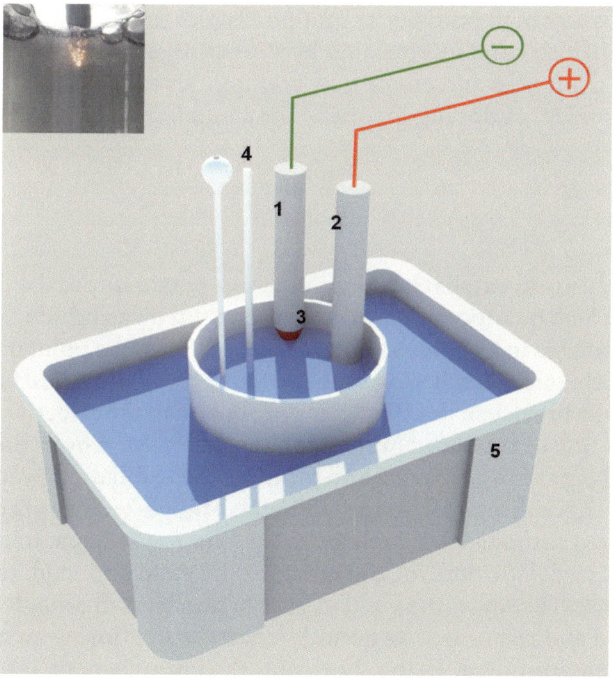

Figure 7.
Experimental setup for production of UPEEG: (1) cathode, (2) anode, (3) vapor plasma envelope, (4) ultrasonic cleaner; inset is image of cathodic electrochemical discharge phenomenon [15].

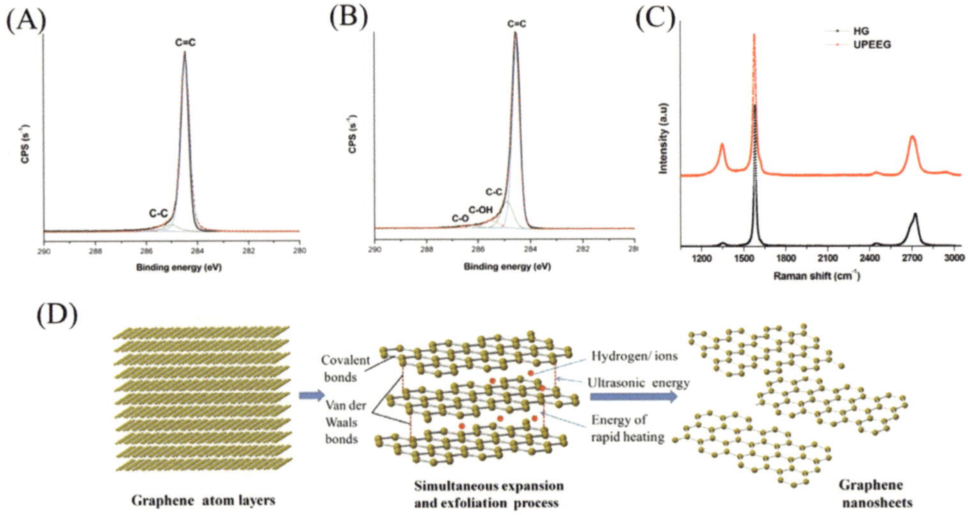

Figure 8.
XPS C1s spectra of (A) HG and (B) UPEEG; (C) Raman spectra of HG and UPEEG; (D) proposed mechanism for the formation of graphene sheets [15].

the main reasons for exfoliation of expanded layers. At first, the shear and shock wave in ultrasonic exfoliating fluid and the thermal expanded or cracked surface are engendered. Subsequently, the cavitation bubbles formed during sonication prompted the disturbance and pulling at the expanded and cracked points in the exfoliating fluid media. Next, the enlarged shear forces appear between neighboring layers and weaken van der Waals forces between layers. Finally, the expanded positions of the adjacent layers are penetrated and intercalated by the hydrogen atoms or ions, which consequently causes the breaking at these points and results in graphene exfoliation. Due to its high porosity of graphite lattice under high diffusion temperature, it can be diffused and intercalated by the hydrogen atoms. We think that this facile method will pioneer and create new possibilities for production of graphene sheets, graphite oxide, or doping graphene for a number of applications.

4. Conclusions

Sonoelectrochemical exfoliation of graphite is a potentially scalable way to achieve high quality and high quantity of graphene production by using simple equipment that is almost available in all chemistry laboratories. From the point of technology view, the electrochemical exfoliation with the aid of ultrasonication is dual benefited not only to easy-to-break van der Waals forces but also to the effective inhibition of the agglomeration of produced graphene nanosheets. Moreover, the presence of ultrasonication herein can accelerate and promote the exfoliation process.

In this chapter, we discussed on the synthesis of high-quality graphene by direct ultrasonic-assisted cathodic electrochemical exfoliation process. Immense works have been done in order to control and enhance the yield and oxidation degree of graphene sheets. However, the yield of graphene sheets in a single layer is still relatively limited and requires long periods of sonication time with massive heat release into the environment. In the short run, the improvement and design optimization on reproducible protocol should be devoted to define the best condition for sonoelectrochemical exfoliation method. Besides, one might seem very presumably that the unreasonable reaction should be existed that taking into account for the formation and exfoliation of graphene nanosheets and their functional groups containing oxygen on the surface. The mechanism is still in its infancy and should be unveiled in the near future to pave the way for controlling of graphene production and its derivatives. Therefore, we hope this chapter will contribute critical insights for understanding the graphene synthesis and hence can open an alternative approach toward 2D materials in general and graphene material in specific.

Author details

Nguyen Van Truong[1], Nguyen Quoc Dung[2], Nguyen Nhat Huy[3], Pham Van Hao[4] and Dang Van Thanh[5]*

1 Faculty of Fundamental Science, Thai Nguyen University of Technology, Vietnam

2 Department of Chemistry, Thai Nguyen University of Education, Vietnam

3 Faculty of Environment and Natural Resources, Ho Chi Minh City University of Technology, VNU-HCM, Vietnam

4 Faculty of Basic Science, TNU-University of Information and Communication Technology, Vietnam

5 Faculty of Basic Science, TNU-University of Medicine and Pharmacy, Vietnam

*Address all correspondence to: thanhdv@tnmc.edu.vn

References

[1] Yi M, Shen Z. A review on mechanical exfoliation for the scalable production of graphene. Journal of Materials Chemistry A. 2015;**3**:11700-11715. DOI: 10.1039/c5ta00252d

[2] Li X, Yu J, Wageh S, Al-Ghamdi AA, Xie J. Graphene in photocatalysis: A review. Small. 2016;**12**:6640-6696. DOI: 10.1002/smll.201600382

[3] Bonaccorso F, Colombo L, Yu G, Stoller M, Tozzini V, Ferrari AC, et al. Graphene, related two-dimensional crystals, and hybrid systems for energy conversion and storage. Science. 2015;**347**: 1246501-1246509. DOI: 10.1126/science.1246501

[4] Liu L, Corma A. Metal catalysts for heterogeneous catalysis: From single atoms to nanoclusters and nanoparticles. Chemical Reviews. 2018;**118**:4981-5079. DOI: 10.1021/acs.chemrev.7b00776

[5] Hummers WS, Offeman RE. Preparation of graphitic oxide. Journal of the American Chemical Society. 1958;**80**:1339. DOI: 10.1021/ja01539a017

[6] Su CY, Lu AY, Xu Y, Chen FR, Khlobystov AN, Li LJ. High-quality thin graphene films from fast electrochemical exfoliation. ACS Nano. 2011;**5**:2332-2339. DOI: 10.1021/nn200025p

[7] Compton RG, Eklund JC, Marken F. Sonoelectrochemical processes: A review. Electroanalysis. 1997;**9**:509-522. DOI: 10.1002/elan.1140090702

[8] Bang JH, Suslick KS. Applications of ultrasound to the synthesis of nanostructured materials. Advanced Materials. 2010;**22**:1039-1059. DOI: 10.1002/adma.200904093

[9] Xu H, Zeiger BW, Suslick KS. Sonochemical synthesis of nanomaterials. Chemical Society Reviews. 2013;**42**:2555-2567. DOI: 10.1039/c2cs35282f

[10] Peller J, Wiest O, Kamat PV. Sonolysis of 2,4-dichlorophenoxyacetic acid in aqueous solutions. Evidence for OH-radical-mediated degradation. The Journal of Physical Chemistry. A. 2002;**105**:3176-3181. DOI: 10.1021/jp003478y

[11] Suslick KS, Price GJ. Applications of Ultrasound to Materials Chemistry. Annual Review of Materials Science. 1999;**29**:295-326. DOI: 10.1146/annurev.matsci.29.1.295

[12] Parvez K, Wu ZS, Li R, Liu X, Graf R, Feng X, et al. Exfoliation of graphite into graphene in aqueous solutions of inorganic salts. Journal of the American Chemical Society. 2014;**136**:6083-6091. DOI: 10.1021/ja5017156

[13] Liu J, Poh CK, Zhan D, Lai L, Lim SH, Wang L, et al. Improved synthesis of graphene flakes from the multiple electrochemical exfoliation of graphite rod. Nano Energy. 2013;**2**:377-386. DOI: 10.1016/j.nanoen.2012.11.003

[14] Lee H, Bratescu MA, Ueno T, Saito N. Solution plasma exfoliation of graphene flakes from graphite electrodes. RSC Advances. 2014;**4**:51758-51765. DOI: 10.1039/c4ra03253e

[15] Van Thanh D, Oanh PP, Huong DT, Le PH. Ultrasonic-assisted cathodic electrochemical discharge for graphene synthesis. Ultrasonics Sonochemistry. 2017;**34**:978-983. DOI: 10.1016/j.ultsonch.2016.07.025

[16] Yu P, Lowe SE, Simon GP, Zhong YL. Electrochemical exfoliation of graphite and production of functional graphene. Current Opinion

in Colloid & Interface Science. 2015;**20**:329-338. DOI: 10.1016/j.cocis.2015.10.007

[17] Chang LC, Hsieh YC, Chen YM, Wu PW, Lee JF. Fabrication of graphene by electrochemical exfoliation in alkaline electrolytes. ECS Transactions. 2014;**58**:33-38. DOI: 10.1149/05824.0033ecst

[18] Wang G, Wang B, Park J, Wang Y, Sun B, Yao J. Highly efficient and large-scale synthesis of graphene by electrolytic exfoliation. Carbon. 2009;**47**:3242-3246. DOI: 10.1016/j.carbon.2009.07.040

[19] Joo EH, Kuila T, Kim NH, Lee JH, Kim SA, Park EG, et al. Electrochemically preparation of functionalized graphene using sodium dodecyl benzene sulfonate (SDBS). Advanced Materials Research. 2013;**747**:246-249. DOI: 10.4028/www.scientific.net/AMR.747.246

[20] Alanyalıoğlu M, Segura JJ, Oró-Solè J, Casañ-Pastor N. The synthesis of graphene sheets with controlled thickness and order using surfactant-assisted electrochemical processes. Carbon. 2012;**50**:142-152. DOI: 10.1016/j.carbon.2011.07.064

[21] Wang J, Manga KK, Bao Q, Loh KP. High-yield synthesis of few-layer graphene flakes through electrochemical expansion of graphite in propylene carbonate electrolyte. Journal of the American Chemical Society. 2011;**133**:8888-8891. DOI: 10.1021/ja203725d

[22] Yang Y, Ji X, Yang X, Wang C, Song W, Chen Q, et al. Electrochemically triggered graphene sheets through cathodic exfoliation for lithium ion batteries anodes. RSC Advances. 2013;**3**:16130-16135. DOI: 10.1039/c3ra43010c

[23] Abdelkader AM, Kinloch IA, Dryfe RAW. Continuous electrochemical exfoliation of micrometer-sized graphene using synergistic ion intercalations and organic solvents. ACS Applied Materials & Interfaces. 2014;**6**:1632-1639. DOI: 10.1021/am404497n

[24] Lei H, Tu J, Yu Z, Jiao S. Exfoliation mechanism of graphite cathode in ionic liquids. ACS Applied Materials & Interfaces. 2017;**9**:36702-36707. DOI: 10.1021/acsami.7b03306

[25] Van Thanh D, Chen H-C, Li L-J, Chu C-W, Wei K-H. Plasma electrolysis allows the facile and efficient production of graphite oxide from recycled graphite. RSC Advances. 2013;**3**:17402. DOI: 10.1039/c3ra43084g

[26] Van Thanh D, Li L, Chu C, Yen P, Wei K. Plasma-assisted electrochemical exfoliation of graphite for rapid production of graphene sheets. RSC Advances. 2014;**4**:6946. DOI: 10.1039/c3ra46807k

[27] Aliofkhazraei M, Rouhaghdam AS, Gupta P. Nano-fabrication by cathodic plasma electrolysis. Critical Reviews in Solid State and Materials Sciences. 2011;**36**:174-190. DOI: 10.1080/10408436.2011.593269

[28] Gupta P, Tenhundfeld G, Daigle EO, Ryabkov D. Electrolytic plasma technology: Science and engineering: An overview. Surface and Coating Technology. 2007;**201**:8746-8760. DOI: 10.1016/j.surfcoat.2006.11.023

[29] Yerokhin AL, Nie X, Leyland A, Matthews A, Dowey SJ. Plasma electrolysis for surface engineering. Surface and Coating Technology. 1999;**122**:73-93. DOI: 10.1016/S0257-8972(99)00441-7

[30] Gupta P, Tenhundfeld G, Daigle EO, Schilling PJ. Synthesis and characterization of hard metal coatings by electro-plasma technology.

Surface and Coatings Technology. 2005;**200**:1587-1594. DOI: 10.1016/j.surfcoat.2005.08.030

[31] Nie X, Tsotsos C, Wilson A, Yerokhin AL, Leyland A, Matthews A. Duplex surface treatments combining plasma electrolytic nitrocarburising and plasma-immersion ion-assisted deposition. Surface and Coatings Technology. 2001;**139**:135-142. DOI: 10.1016/S0257-8972(01)01025-8

[32] Paulmier T, Bell JM, Fredericks PM. Deposition of nano-crystalline graphite films by cathodic plasma electrolysis. Thin Solid Films. 2007;**515**:2926-2934. DOI: 10.1016/j.tsf.2006.08.027

[33] Campos CS, Spada ER, De Paula FR, Reis FT, Faria RM, Sartorelli ML. Raman and XRD study on brookite-anatase coexistence in cathodic electrosynthesized titania. Journal of Raman Specroscopy. 2012;**43**:433-438. DOI: 10.1002/jrs.3048

Section 2

Sonochemical Applications

Chapter 5

Sonochemistry: Applications in Biotechnology

Lingayya Hiremath, S. Nipun, O. Sruti, N.G. Kala and B.M. Aishwarya

Abstract

Sonochemistry is a branch dealing with effects of chemical as well as sound wave as the name suggest. The sound waves are ultrasonic, i.e., high frequency waves (20 kHz can extent to 10 MHz and above) beyond the range of a human ear (20–20 kHz). Sonochemistry technology is incorporated into both mechanistic and synthetic studies. An important event called acoustic cavitation take place where microbubbles grow and under the influence of ultrasonic waves they collapse. Sonoluminescence is one of the outcomes of cavitation which leads to homogeneous sonochemistry. Sonochemistry has also entered one of the major developing field biotechnology from basic activation of enzyme to preparation of catalyst. It is also used for the fabrication of nanomaterial which comes under the liquid phase method. One disadvantage of nanomaterial preparation is the amount of time it consumes to show results. This can be eliminated when biotechnological research is conducted in conjunction with sonochemical application. Latest research results have proved that ultrasound irradiation is both time and cost-effective approach for any bio-processes like enhancement of emulsification and trans-esterification of fatty acids for bio-fuel products. Bio-process monitoring and dewatering of sludge have also been accelerated. This chapter contains introductory information on sonochemistry.

Keywords: sonochemistry, acoustic cavitation, nanoparticles, collides, medicine, water treatment, sonocatalyst

1. Introduction to sonochemistry

1.1 About sonochemistry

We are all familiar with the terms chemistry and sound when these two terms are put together into practice, a new field is developed called sonochemistry. This is quite new to us and is still a developing area of science. In simple words this field involves the study of effect of sound in chemical reactions in a given solution caused by acoustic cavitation [1, 2]. The basis of sonochemistry is use of lower range of ultrasound with higher power that causes significant physical and chemical changes. First commercial use of ultrasound dates back into 1917 by Langevin to measure the depth of water [1]. In 1927, Robert Williams Wood (1868–1955) and Alfred Lee Loomis (1887–1975) conducted an experiment to demonstrate effect of ultrasound in water [3]. Later in 1960s use of ultrasound entered industries

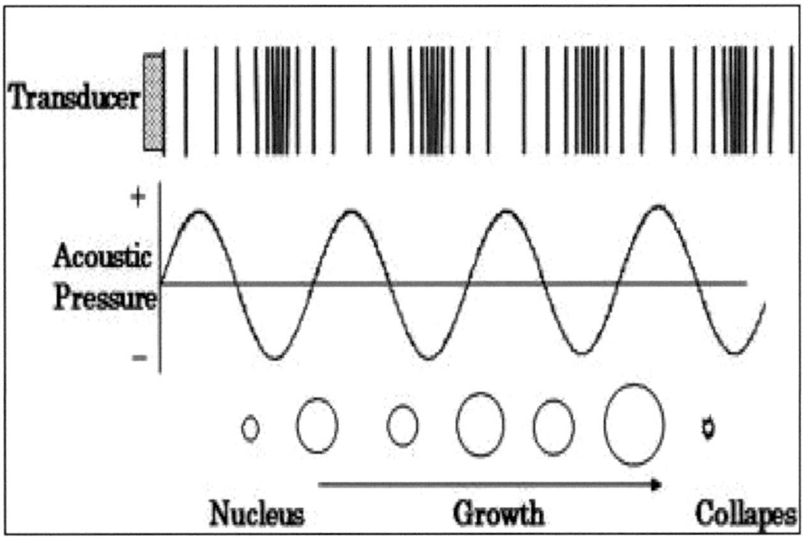

Figure 1.
Acoustic cavitation caused by ultrasound.

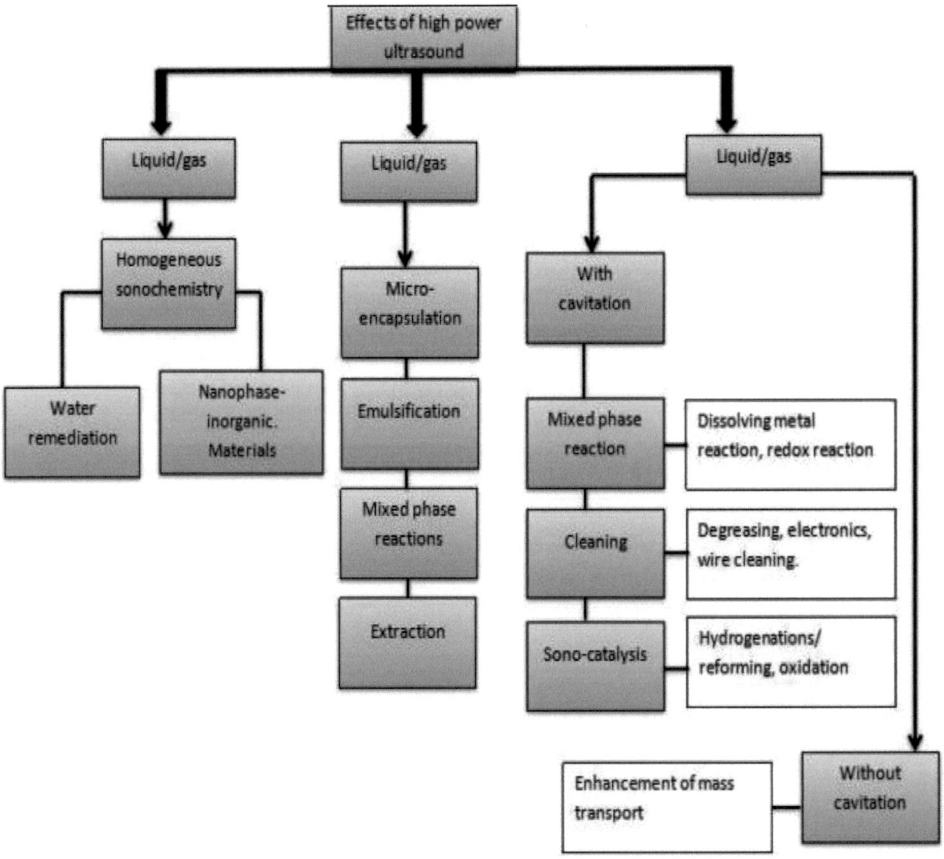

Figure 2.
Chemical and physical effects of ultrasound [4].

and soon became a means to accomplish advancement using technology. These methods were used in many chemical processes to either initiate or enhance the on-going reactions.

Now that we are familiar with the topic next question is how sound effects the chemical reaction in a solution they are introduced into. Acoustic cavitation is responsible for the enhancement of these processes. When sound waves (ultrasound) are passed into solutions they generate mechanical vibrations along with acoustic streaming. These solutions in normal conditions contain dissolved gasses nuclei, which collapse under ultrasound field. This cavitation (oscillation and collapse) results in many physical changes like shockwaves, shear forces, turbulence, etc., along with physical changes it is an adiabatic process and generates very high temperature inside the bubbles for a short interval (**Figure 1**) [1, 2, 4].

1.2 Effects of sonochemistry

These are both chemical and physical effects in which chemical falls under homogeneous sonochemistry of liquids, heterogeneous sonochemistry of liquid-liquid or liquid-solid systems, and sonocatalysis. Based on earlier studies, effects of ultrasound on slurries of inorganic solids are shown (**Figure 2**).

2. Application of sonochemistry

2.1 Nanostructured inorganic materials

Over the past few years sonochemical reactions have been chosen for a general approach towards the synthesis of nanophase materials. Due to distinct behaviour of nanosized material compared to the bulkier ones [5]. These small clusters have electronic structures with high density. Both gas phase and liquid phase techniques are used to synthesis them. With these different phase techniques and also their combination, the sonochemical approach is included [4, 5]. Various forms of nanophase material are generated by simply changing the reaction medium as shown (**Figure 3**).

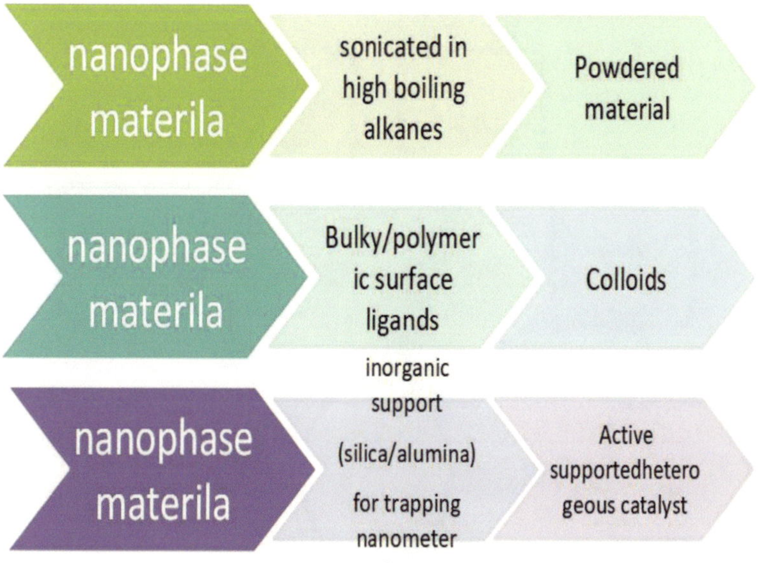

Figure 3.
Different forms of nanomaterial generated by changing reaction medium of the phase.

2.1.1 Different types of nano products obtained

Nanomaterials obtained using sonochemistry are of various forms. Different forms exist based on their usage. As an matter of interest, nano-sized particles serve as great source for action. This is the reason they are so popular. In **Figure 3** most common nano-product are listed. These materials are formed using different techniques and processes. Sometimes combinations of techniques are used in conjunction with sonochemistry. Also due to their diverse nature they consume different interval of time to become fully functional. Based on these categories the options are narrowed for their usage. Some of these materials are listed in **Figure 4** [4, 6].

2.1.1.1 Amorphous metals

A dull black powder was obtained by sonication of iron pentacarbonyl in decane at 0°C under continuous flow of argon. Amorphous nature was verified by different techniques like SEM, different scanning calorimetry, electron micro-diffraction, X-ray powder diffraction and neutron diffractions were used. EMD revealed a diffuse ring pattern (amorphous characteristic). The amorphous metals are formed as a result of extremely high cooling rate during acoustic cavitation. This was shown by DSC (noted transition temperature is 308°C). The size of the particles were in the range of 4–6 nm [4].

2.1.1.2 Alloys

The compositions of Fe/Co alloy can be controlled by changing the ratio of solution concentrations of both the individuals. They are readily available and are thermally stable at adequate bulk solution temperature. The solid-solution nature is determined by energy dispersive X-ray measurements. The products are homogeneous on a nano-scale. They are initially amorphous and after heat treatment at 400°C under H_2 gas for 2 h (**Figure 5**).

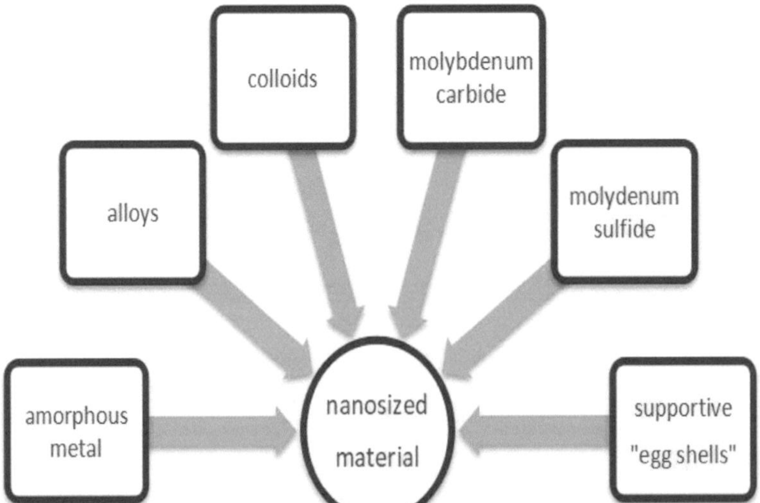

Figure 4.
Most common nanophase materials produced using sonochemical reactions.

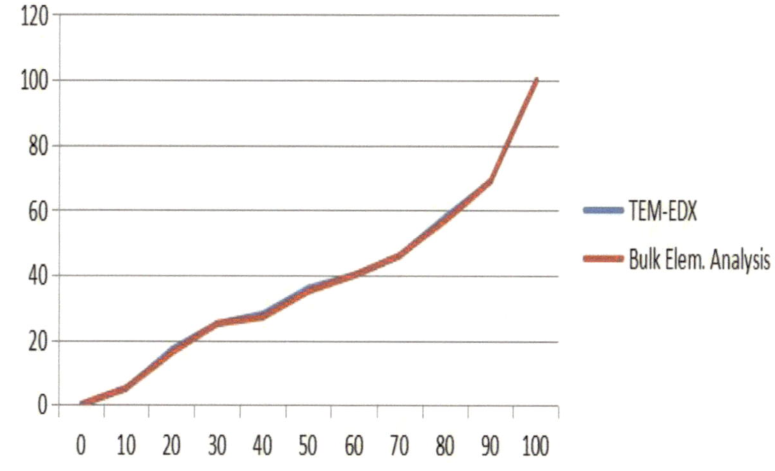

Figure 5.
Bulk and EDX analysis of Fe/Co alloy prepared sonochemically from respective precursors [4].

2.1.1.3 Colloids

Colloids of ferromagnetic materials are made by exhaustive grinding of magnetite (Fe_3O_4) in ball or vibratory mills for many weeks. Wide range of particles (based on size) is produces due to presence of surfactants. A new method is developed using high intensity ultrasound to produce stable colloids of iron. In this method volatile organo-metallic compounds are decomposed sonochemically. In presence of polyvinylpyrrolidone, the size of colloidal iron particles are in range of 3–8 nm whereas in presence of oleic acid they are in uniformly distributed at 8 nm [4] (**Figure 6**).

2.1.1.4 Molybdenum carbide

Sonochemical decomposition of molybdenum hexacarbonyl in hexadecane by ultrasound irradiation produces a black powder. When this powder is heated for

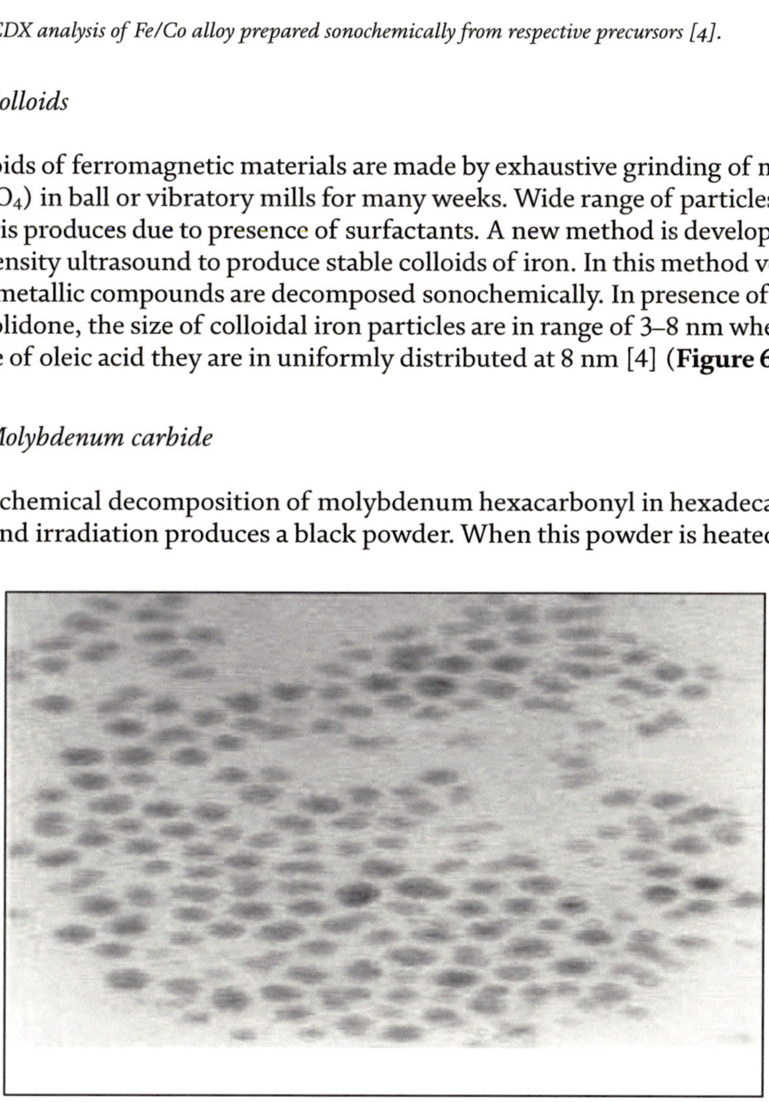

Figure 6.
Transmission electronic micro-graphic view of sonochemically prepared iron colloid (stabilized by oleic acid) [7].

12 h under He flow at 450°C. SEM revealed the extremely porous surface and TEM confirmed the size of particles of the porous aggregate, i.e., 3 nm diameter [4].

2.1.1.5 Molybdenum sulfide

Due to its layered structure, MoS_2 is also called as the standard automotive lubricant. It is prepared by high intensity ultrasound irradiation of molybdenum hexacarbonyl and sulfur in 1,2,3,5-tetramethylbenzene. Under the flow of He the amorphous product was heated at 450°C to get crystallized MoS_2. The sonochemically prepared MoS_2 has much greater edge and defect content than conventionally prepared ones as the layers must bend, break or otherwise distort to fit outer surface of the 15 nm particle size [4] (**Figure 7**).

2.1.1.6 Supportive "egg shells"

When eggshells with uniform-sized nanoparticles of metals are deposited on the outer surface of supports, potential advantages for catalyst preparation increases greatly compared to non-uniform particles. Ultrasonic irradiation of decane solutions of iron pentacarbonyl, $Fe(CO)_5$, in the presence of silica gel produces a silica-supported amorphous nanostructured iron-iron particles on the outer surface of are formed during cavitation. By changing the initial composition of iron precursor, its loading on SiO_2 can be verified. TEM confirmed the high dispersion of iron particles (3–8 nm) produced by sonolysis of $Fe(CO)_5$ on the SiO_2 surface [4].

2.2 Synthesis of bio-materials

Protein micro-spheres are one of the most notable bio-materials. The development of aqueous sonochemistry for bio-materials synthesis is still an on-going

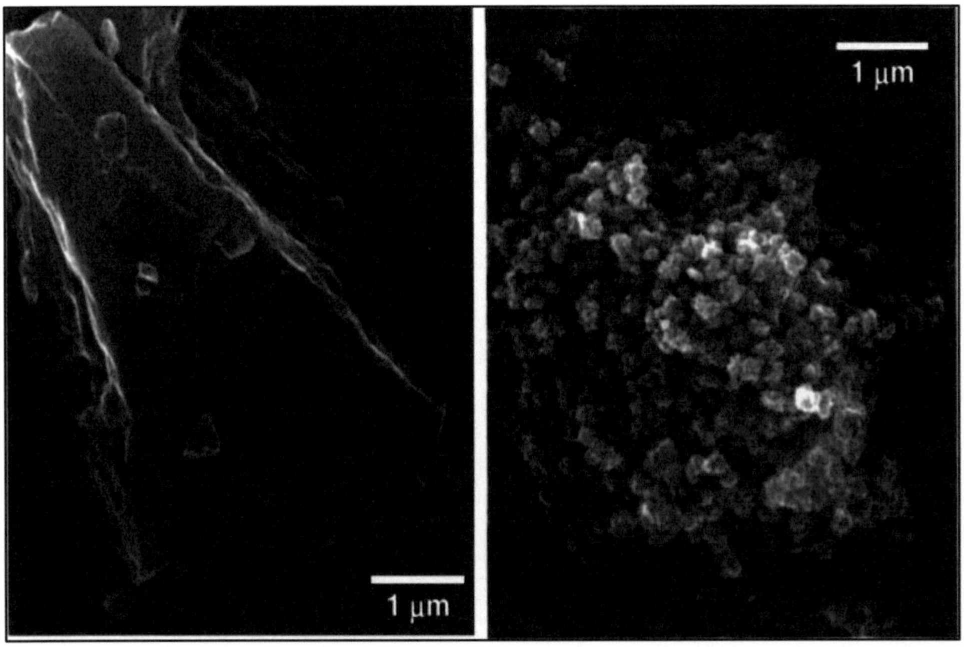

Figure 7.
Conventionally prepared MoS_2 (left) and sonochemically prepared MoS_2 (right) [8].

research area particularly of micro-encapsulation. High intensity ultrasound is used on simple protein solutions to make both air-filled micro bubbles and non-aqueous liquid-filled micro-capsules. Micro-spheres are stable for months and can be intravenously injected to pass unimpeded through the circulatory system due to their smaller size compared to erythrocyte scan. The particle size can be determined as a function of sonication time, suggesting the disintegration of the aggregates on longer exposure to ultrasound radiation [4, 9] (**Figures 8–10**).

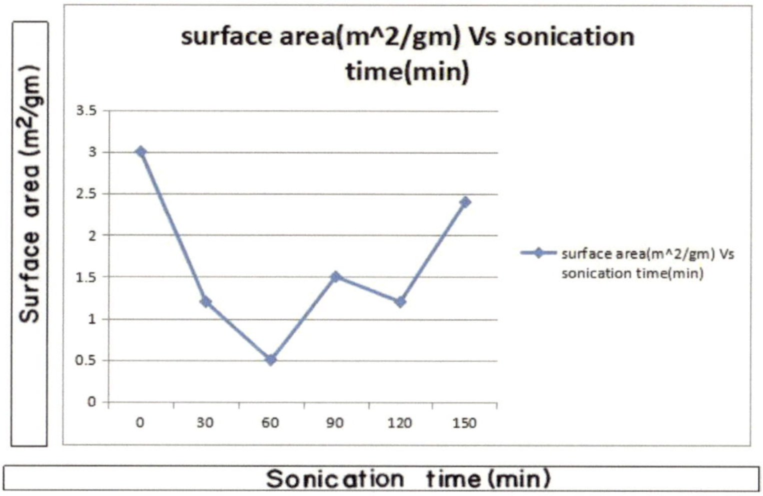

Figure 8.
The variation of surface area of micro-spheres with time is represented in the above graph [4, 9].

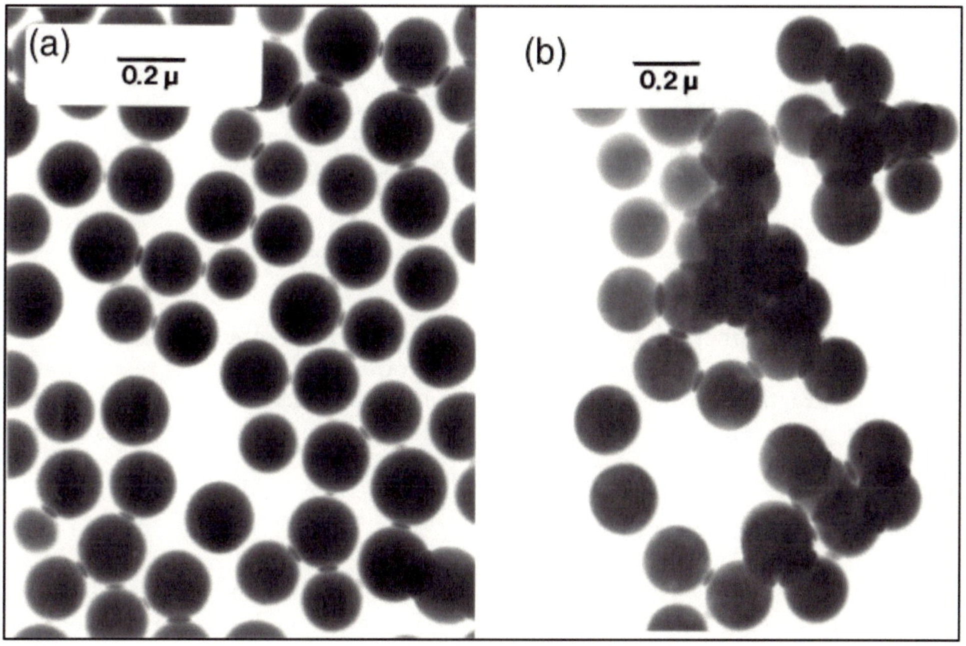

Figure 9.
TEM micro-graphs of (a) as-prepared Stober's silica in the size range of 150–200 nm and (b) the silica micro-spheres irradiated with ultrasound for 2 h in decalin medium [9].

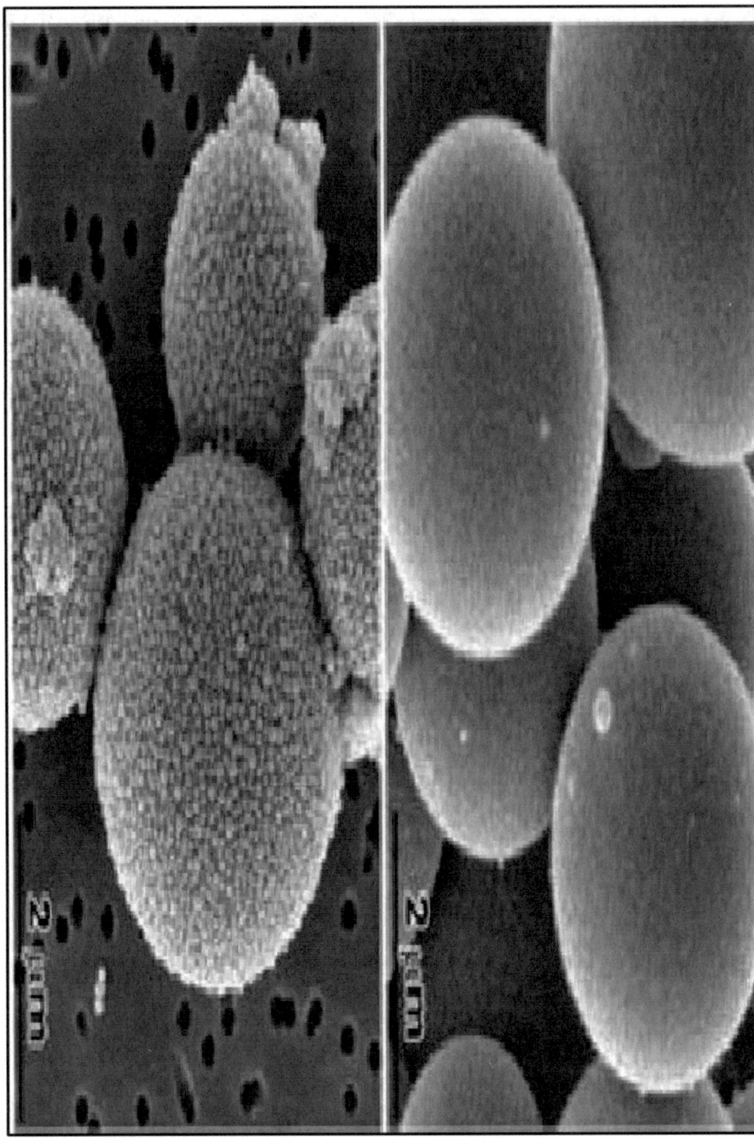

Figure 10.
SEM images of micro-spheres prepared by sonication of bovine serum albumin [10].

Protein micro-spheres have a wide range of biomedical applications. They are also used as a echo contrast agents for sonography [11], magnetic resonance imaging contrast enhancement [12–14], oxygen or drug delivery [15] sonochemically produced hemoglobin micro-bubbles, hollow and solid spheres and micro-spheres.

2.3 Modification of inorganic materials

Ultrasound is widely used to accelerate chemical reactions in liquid-solid heterogeneous systems. This sonochemical enhancement has become a synthetic technique for heterogeneous organic as well as organo-metallic reactions [4, 16, 17]. These reactions mostly include metals like such as Mg, Li and Zn and also apply to reactive inorganic salts and to main group reagents. Less work has been done with unreactive metals.

2.3.1 Surface cavitation

Cavitation near extended liquid-solid interfaces and cavitation in pure liquids, both are very different [4, 17]. Micro-jet impact and shockwave damage are responsible for effects caused by cavitation near surfaces. The special symmetry of the liquid particle in motion during cavity collapse often induces a deformation within the cavity. The potential energy of the expanded bubble is converted into kinetic energy of a liquid jet that extends through the bubble's interior and penetrates the opposite bubble wall. As most of the energy is converted to accelerate jet, the velocity attained is in hundreds of m/s. Such energy concentration can caused extreme damage to the boundary surface (**Figure 11**).

The later one, i.e., cavitation in pure liquid, mechanism of cavitation-induced surface damage invokes shockwaves created by cavity collapse in the liquid [4]. Micro-jets and shockwaves on the surface, creates the localized erosion responsible for ultrasonic cleaning and. These effects can generate newly exposed highly heated surfaces and even eject metal from the surface [18].

2.3.2 Inter-particle collision

For ultrasonic frequencies greater than 20 kHz, damage associated with micro-jet formation cannot occur for solid particles smaller than 200 mm. This takes on a special importance for sonochemistry [19]. This is the reason, they are preferred as catalysts. Passing shock waves over particles in close proximity to one another results in high-velocity inter-particle collisions. Right angled collision can drive metal particles at sufficiently high velocities to cause melting effect at the collision point [4]. Energy generated is determined by the volume of melted region under the

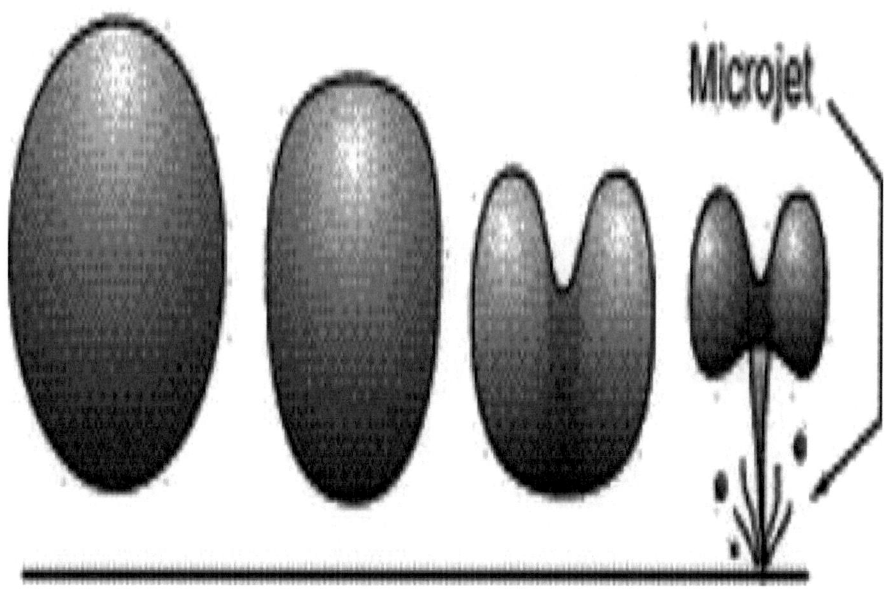

Figure 11.
It shows the collapse of a bubble later expanded to a liquid jet (cavitation near surface).

impact of collision. A lower estimate of the velocity of impact was several hundred m/s, or roughly one half the speed of sound! [4] (**Figures 12** and **13**).

2.3.3 Ultrasound on heterogeneous catalysts

These kinds of catalysts are rare and expensive. Using ultrasound activation of less reactive is less costly. The effect occur in three distinct stages

1. During formation of supported catalysts.

2. Activation of performed catalysts.

3. Enhancement of catalytic behavior during its reaction

Increases in turnover rates were usually observed upon ultrasonic irradiation even though they were rarely more than 10 folds but were appropriate. The hydrogenation of alkenes by Ni powder is enormously enhanced by ultrasonic irradiation [4]. At a macroscopic scale, the initially crystalline surface are agglomerated into small particles by ultrasound irradiation. This effects is due to inter-particle collisions caused by cavitation-induced shockwaves, revealed that there is a striking decrease in the thickness of the oxide coat after ultrasonic irradiation. It is the removal of this passivating layer [4, 7] (**Figure 14**).

2.4 Lignocellulosic biomass

Lignocellulosic biomass is a natural renewable chemical feedstock that is used to produce high value-added chemicals and platform molecules. There is variety of aspects concerning the valorization of lignocellulosic biomass into desirable products. For biomass conversion, few of the existing technologies require high temperatures

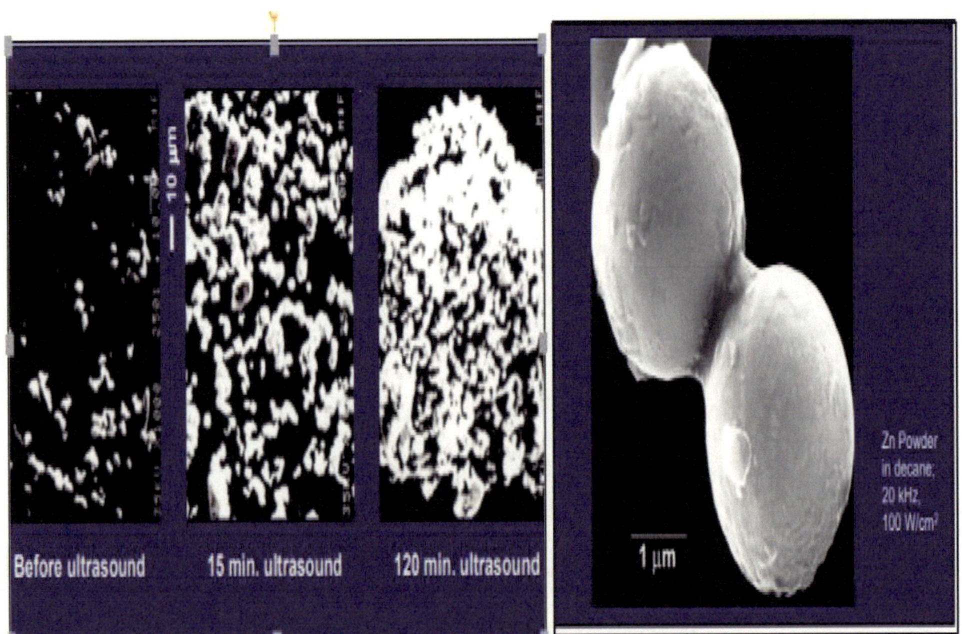

Figure 12.
Cavitation effect on surface morphology (left), inter-particle collision (right).

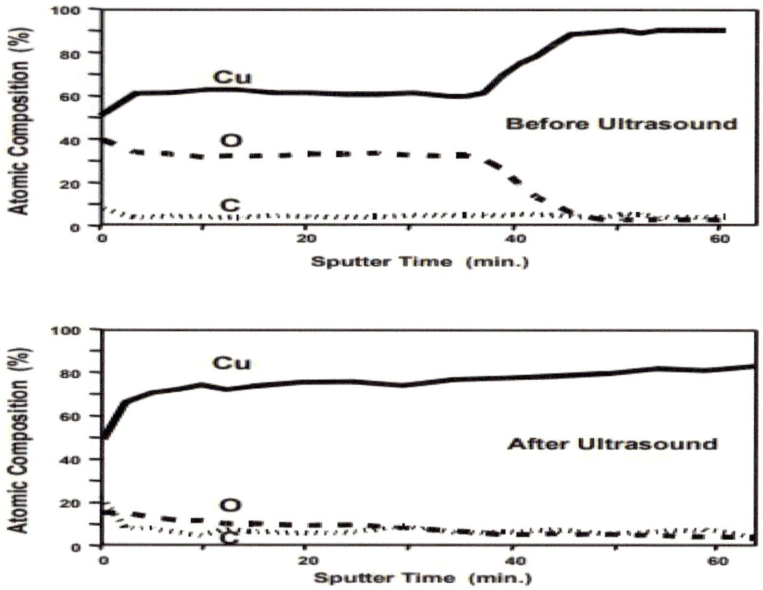

Figure 13.
The effect of ultrasonic irradiation of Cu powder slurries on the surface composition. 50 min. Sputter time is roughly 1 mm depth [4].

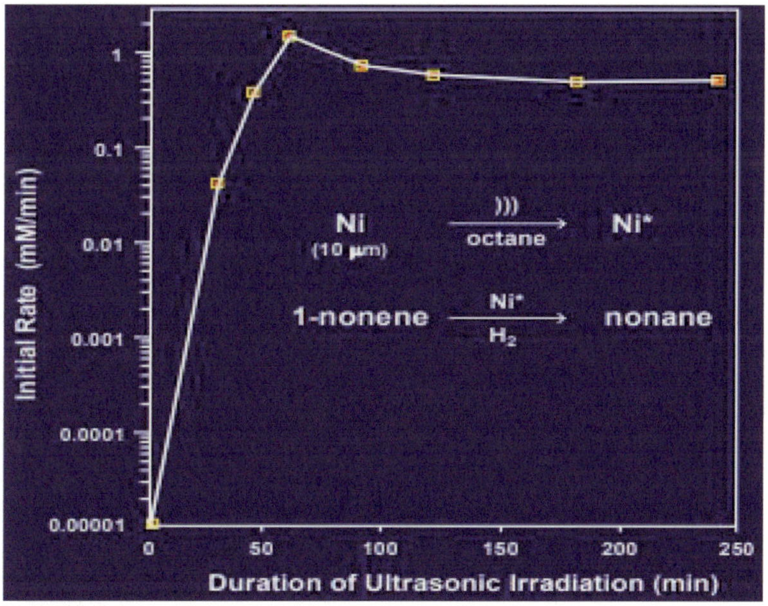

Figure 14.
Sonochemical activation of Ni powder.

and pressures. To fulfill the extreme conditional requirement, key technological innovations based on more economical and environmental methodologies are being explored both in academic laboratories and in industry. Sonocatalyst constitutes a substitute means offering innovative strategies to improve biomass (**Figure 15**).

Heterogeneous sonocatalysis as an emerging advanced oxidation process (AOP), is a potential treatment method for water and has removed persistent organic compounds in the in past few decades [21].

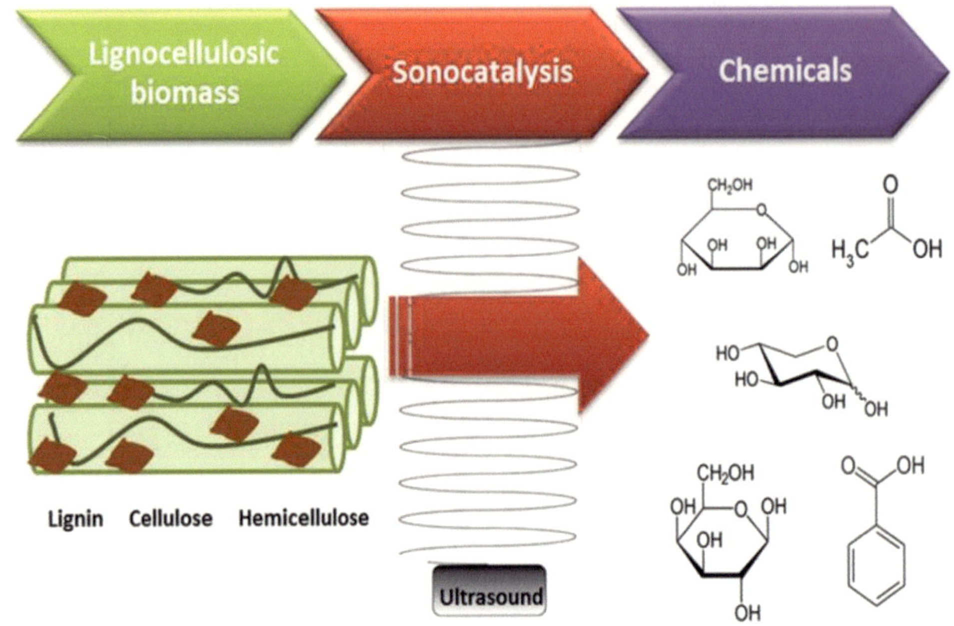

Figure 15.
Biomass valorization [20].

2.5 Sonoluminescence

Sonoluminescence is a phenomenon that occurs when a small gas bubble is acoustically suspended and periodically driven in a liquid solution at ultrasonic frequencies, resulting in bubble collapse, cavitation, and light emission.

Sound waves are aimed at an air bubble trapped in a flask. The sound waves cause the bubble to oscillate furiously

- The bubble starts out at a size around 5 μm.
- It expands to itself up to 50 μm (roughly). At this large size there is a near-vacuum inside the bubble because of the relatively few air molecules present. This low-pressure near-vacuum region is surrounded outside the bubble by a much higher-pressure region.
- A catastrophic collapse of the bubble to between 0.1 and 1 microns occurs.
- During this compression phase a flash of light emerges from the bubble.

The conversion of low energy density sound waves into light requires a concentration of energy by a factor of 1 trillion (**Figure 16**).

2.6 Sludge treatment

Sludge treatment in waste-water treatment plants are one of the most difficult challenges. This new technology of treating water with ultrasound is quite helpful. Ultrasound treatment promote hydrolysis during sludge treatment.

The basic principal is based on the destruction of bacterial cells and organics. Bacterial cells release their contents, which are then available for consumption by

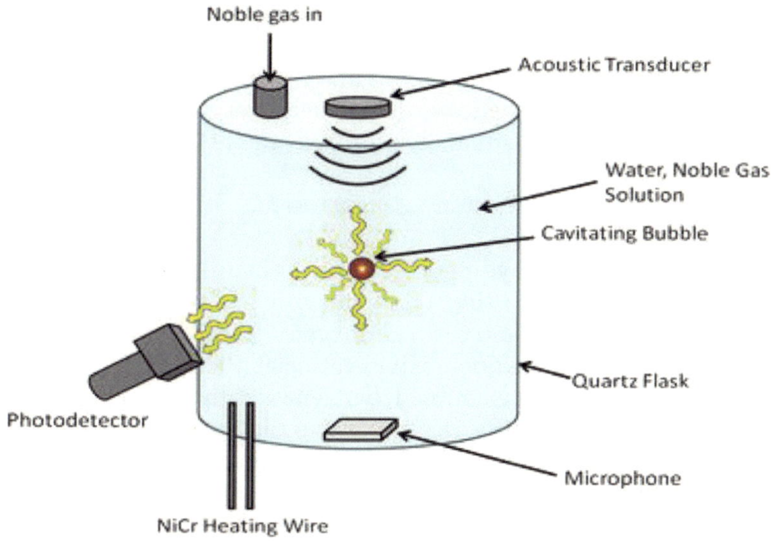

Figure 16.
A setup similar to the following is required to create a bubble that can sonoluminescence.

other species. The organics are broken down into smaller fractions which can be readily biodegradable.

When an ultrasonic field is applied to a liquid medium, microbubbles are produced during the cavitation process. These cavitation bubbles implode under extreme high temperature and pressure. The force of the bubble implosions breaks up sludge particles when applied in sewage sludge treatment. The extent of the intended effects depends on the high or low intensity of the applied ultrasonic field [22].

3. Sonochemistry in market

3.1 Industrial level

When it comes to an industrial level of production, cost efficient and avoiding wasteful reactions is considered primary when we talk about catalysts, maximum efficiency is the main concern and it can vary depending upon the metals or any other precursors used. Heterogeneous catalysts containing two metals often show unusual activity or selectivity for a wide range of industrially important reactions. Due to these reasons, catalytic studies of alloys, e.g., Fe/Co (as mentioned in the applications), was made using sonochemistry. Dehydrogenation of hydrocarbons is the important and desirable reaction, and hydrogenolysis (to methane generally) is a wasteful side-reaction in industries because of its high activity (more than required) as hydrogenolysis catalysts for commercial usage [4].

3.1.1 Preparation of nanostructured alloys

All catalysts were treated under H_2 gas flow at 400°C for 2 h which causes crystallization before the catalytic studies. The catalytic selectivity is shown in **Figure 1**. Two different products were formed in the cyclohexane reaction. The first one is benzene from the dehydrogenation reaction and other one is aliphatic hydrocarbons from the hydrogenolysis. Fe/Co alloys are all active catalysts for cyclohexane conversion the activity is higher for Co and decreases with increase in Fe content. Compared to pure metals, alloys generated much more dehydrogenation products. When 1:1 ratio of alloys was used, it had the highest selectivity along with higher dehydrogenation activity. The sonochemical preparations inevitably produce catalysts with small amounts of surface carbon and hence with high dehydrogenation selectivity compared to less selectivity of conventionally prepared ones [4] (**Figures 15** and **16**).

3.1.2 Preparation of nanostructured molybdenum carbide

To compare the activity, dehydrogenation versus hydrogenolysis of cyclohexane served as the standard reaction using a flow catalytic micro-reactor. Commercial ultrafine powders of platinum and ruthenium were also used remove surface contaminants under identical conditions, after heating at 400°C for 3 h under H_2 flow to. At all reaction temperatures examined, benzene was the only product formed for both samples and their activities were comparable and no hydrogenolysis product was detected. Results showed that for dehydrogenation of alkanes, sonochemically prepared nanostructured molybdenum carbide has electivity similar to Pt rather than to Ru [4, 23] (**Figures 17** and **18**).

3.2 Biofuel production

Ultrasonication has recently received attention as a novel bio-processing tool for process intensification in many areas of downstream processing. Ultrasonic intensification (periodic ultrasonic treatment during the fermentation process) can result in a more effective homogenization of biomass and faster energy and mass transfer to biomass over short time periods which can result in enhanced microbial growth. Ultrasonic intensification can allow the rapid selective extraction of specific biomass components and can enhance product yields which can be of economic benefit [24]. During lipid extraction from biomass, the physical effects of ultrasonication

	Hydrogenolysis	Dehydrogenation
Products	Methane	Benzene
Surface structure	Sensitive	Insensitive
Active site	Ensemble of metal atoms	Single metal atom
Active catalysts	Ru,Os,most metals	Pt

Figure 17.
Catalytic dehydrogenation versus hydrogenolysis.

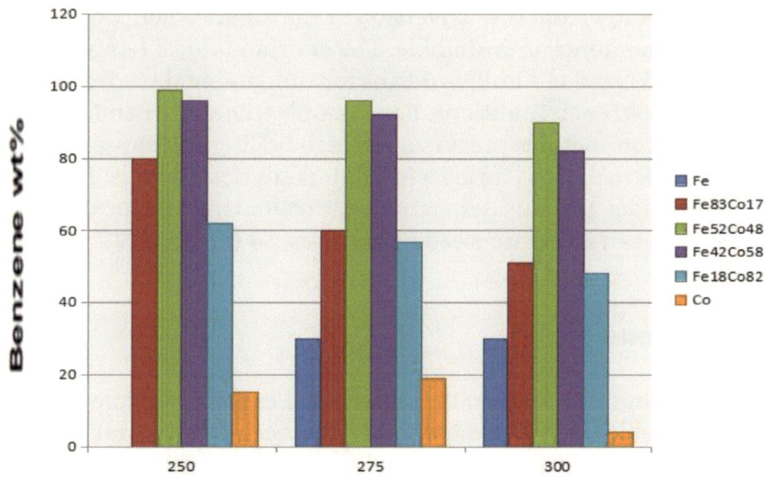

Figure 18.
Selectivity for cyclohexane dehydrogenation (to benzene) over hydrogenolysis (to methane) by sonochemically prepared Fe/Co alloy.

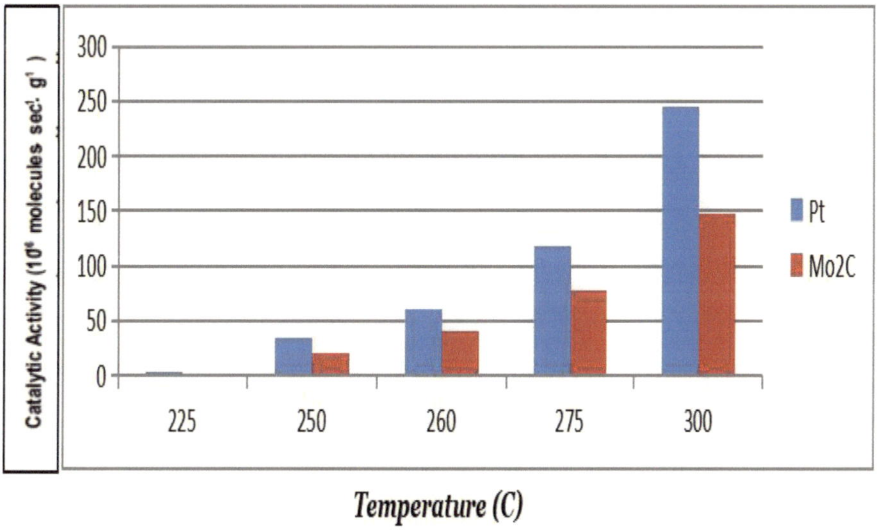

Figure 19.
Catalytic activity of sonochemically prepared Mo_2C compared to Pt.

can significantly enhance the lipid yield. Micro-turbulence can lead to a more efficient mixing of the biomass and solvent (without induction of shear stress), while shock waves can cause rupture of the cell wall. Ultrasound can also generate intense local turbulence in the medium, pushing the extracted lipids away from the surface of the microbial cells, and thus, maintaining a constant concentration gradient for continuous diffusion of lipids from the cells [24] (**Figure 19**).

4. Conclusion

A diverse set of applications of ultrasound to enhancing chemical reactivity has been explored, with important applications in mixed phase synthesis, materials chemistry, and biomedical uses. Bubble collapse in liquids results in an enormous concentration of energy from the conversion of the kinetic energy of liquid motion into heating of the contents of the bubble. The enormous local temperatures and pressures so created provide a unique means for fundamental studies of chemistry and physics under extreme conditions. For example, the sonochemical decomposition of volatile organo-metallic precursors in high boiling solvents produces nanostructured materials in various forms with high catalytic activities. Nanostructured metals, alloys, carbides and sulfides, nanometer colloids, and nanostructured supported catalysts can all be prepared by this general route.

Acknowledgements

This work was supported by our respected teacher and Principle R V C E. We would like to acknowledge the guidance and help received from our college. The studies of the effects of ultrasound on slurries were primarily that of D.J. Casadonte and S.J. Doktycz; and early catalytic studies were by A.A. Cichowlas. We also thank respective authors of the journals we referred to gather a large sum of information.

Author details

Lingayya Hiremath*, S. Nipun, O. Sruti, N.G. Kala and B.M. Aishwarya
Department of Biotechnology, RV College of Engineering, Autonomous Institution Affiliated to Visvesvaraya Technological University, Belagavi, Bengaluru, Karnataka, India

*Address all correspondence to: lingayah@rvce.edu.in

References

[1] Mason TJ. Acoustic cavitation and sonoluminescence. Philosophical Transactions: Mathematical, Physical and Engineering Sciences. 1999;**357**(1751):355-369

[2] Ashokkumar M, Mason T. "Sonochemistry", Kirk-Othmer Encylcopedia of Chemical Technology. John Wiley and Sons; 2007. DOI: 10.1002/0471238961.1915141519211912.A01.PUB2 [Accessed: 19 October 2007]

[3] Dieke GH. Robert Williams Wood 1868-1955. Biographical Memoirs of Fellows of the Royal Society. 1956;**2**:326-345. DOI: 10.1098/rsbm.1956.0022

[4] Suslick KS, Fang MM, Hyeon T, Mdleleni MM. Applications of sonochemistry to materials synthesis. In: Crum LA, Mason TJ, Reisse J, Suslick KS, editors. Sonochemistry and Sonoluminescence. Dordrecht, Netherlands: Kluwer Publishers; 1999. pp. 291-320

[5] Mason TJ, Petrier C, Parsons S. Advanced Oxidation Processes for Water and Wastewater Treatment. London,UK: IWA Publishing; 2004. pp. 185-208

[6] Ashokkumar M. The characterization of acoustic cavitation bubbles—An overview. Ultrasonics Sonochemistry. 2011;**18**(4):864-872

[7] Journal of the American Chemical Society. 1996;**118**:11960-11961

[8] Price GJ. In: Suslick KS, editor. Application of Ultrasound to Materials Chemistry, Urbana, Illinois, USA: Department of Chemistry, University of Illinois at Urbana-Champaign; 1999

[9] Ramesh S, Koltypin Y, Gedanken A. Ultrasound driven aggregation and surface silanol modification in amorphous silica microspheres. Department of Chemistry, Bar-Ilan University, Ramat-Gan 52900, Israel

[10] 231798162_Ultrasound_Driven_Aggregation_and_Surface_Silanol_Modification_in_Amorphous_Silica_Microspheres

[11] Wang Y, Zeiger BW, Suslick KS. Sonochemical synthesis of nanomaterials. Chemical Society Reviews. 2013;**42**(7):2555-2567. DOI: 10.1039/C2CS35282F

[12] Keller MW, Feinstein SB. In: Kerber RE, editor. Echocardiography in Coronary Artery Disease. New York: Future; 1988

[13] Liu KJ, Grinstaff MW, Jiang J, Suslick KS, Swartz HM, Wang W. Development of biocompatible implants of fusinite for in vivo EPR oximetry. Biophysical Journal. 1994;**67**:896-901

[14] Eckburg JJ, Chato JC, Liu KJ, Grinstaff MW, Swartz HM, Suslick KS, et al. The measurement of temperature with electron paramagnetic resonance spectroscopy. Journal of Biomechanical Engineering. 1996;**118**:193-200

[15] Webb AG, Wong M, Kolbeck KJ, Magin RL, Wilmes LJ, Suslick KS. Handbook on applications of ultrasound: Sonochemistry for sustainability. Journal of Magnetic Resonance Imaging. 1996;**6**:675-683

[16] Wong M, Suslick KS. In: Suslick KS, Fang MM, Hyeon T, Mdleleni MM, editors. Applications of Sonochemistry to Materials Synthesis. Urbana, USA: Department of Chemistry University of Illinois at Urbana-Champaign; 1995

[17] Suslick KS. Sonochemistry. In: Kirk-Othmer Encyclopedia of Chemical Technology. 4th ed. Vol. 26. New York: John Wiley & Sons; 1998. pp. 517-541

[18] Luche J-L. Comptes Rendus Series IIB 323, 203 and 337; 1996

[19] Preece CM, Hansson IL. Sonochemistry and sonoluminescence. Advances in the Mechanics and Physics of Surfaces. 1981;**1**:199

[20] Amso NN. Applications of Therapeutic Ultrasound in Medicine. Ultrasonics Sonochemistry, Elsevier; 1994

[21] Athawale AA, Katre PP, Kumar M, Majumdar MB. Synthesis of CTAB-IPA reduced copper nanoparticles. Materials Chemistry and Physics. 2005;**91**:507-512

[22] Barbera WP. Ultrasound improves sewage sludge treatment

[23] Celebrating 25 Years of Ultrasonics Sonochemistry. Ultrasonics Sonochemistry, Elsevier

[24] Naveena N, Armshaw P, Tony Pembroke J. NCBI, Ultrasonic Intensification as a Tool for Enhanced Microbial Biofuel Yields. Limerick, Ireland: Molecular Biochemistry Laboratory, Materials and Surface Science Institute, Department of Chemical and Environmental Sciences, University of Limerick

Chapter 6

Ultrasound Technology Integration into Drinking Water Treatment Train

Raed A. Al-Juboori and Les Bowtell

Abstract

Fresh water is one of the main sources for drinking water production. Due to increasing contamination caused by extreme weather events such as flood and drought as well as urbanization activities, the quality of this source continues to deteriorate. In order to maintain producing high-quality water from heavily contaminated sources, more chemicals are added to water in conventional treatment plants. This practice generates serious health problems such as the formation of disinfection by-products (DBPs) and the increase of coagulants residues (e.g., Al) in the treated water. Combining chemical-free techniques with conventional treatment processes can be a potential solution for such problems. When evaluating various techniques, ultrasound appears to be a sensible choice for improving contaminants removal from surface water. This chapter sheds light on the exacerbating problem of fresh water contamination and succinctly reviews chemical-free techniques' options for water treatment. The focus of this chapter is directed toward providing critical and insightful discussion of fundamentals, mechanisms, and reaction pathways of ultrasound technology for water treatment application. Recommendations for the best location and operating settings of ultrasound application in conventional water treatment train will be provided based on energy saving and minimal downstream impact criteria.

Keywords: ultrasound technology, pulse mode, square wave, dissolved organic carbon, coagulation, filtration and disinfection

1. Common challenges in conventional drinking water treatment systems

Water is an essential element for living systems. It facilitates the transport of nutrients and waste products within the body of living creatures [1]. Surface water is one of the important supplies for drinking water production [2]. Recently, surface water has been increasingly contaminated by microorganisms, organic matter, particles, and solids due to the developing effects of human activities and climate change as is depicted in **Figure 1** [3–5]. This increase in the concentration of surface water contaminants has led to the increase in the cost associated with the treatment of water. The quality of the produced water has also deteriorated as a result of increased contamination. According to the World Health Organization (WHO), 5 million death cases per year worldwide are caused by poor quality drinking water [6]. These problems have made the enhancement of surface water treatment to

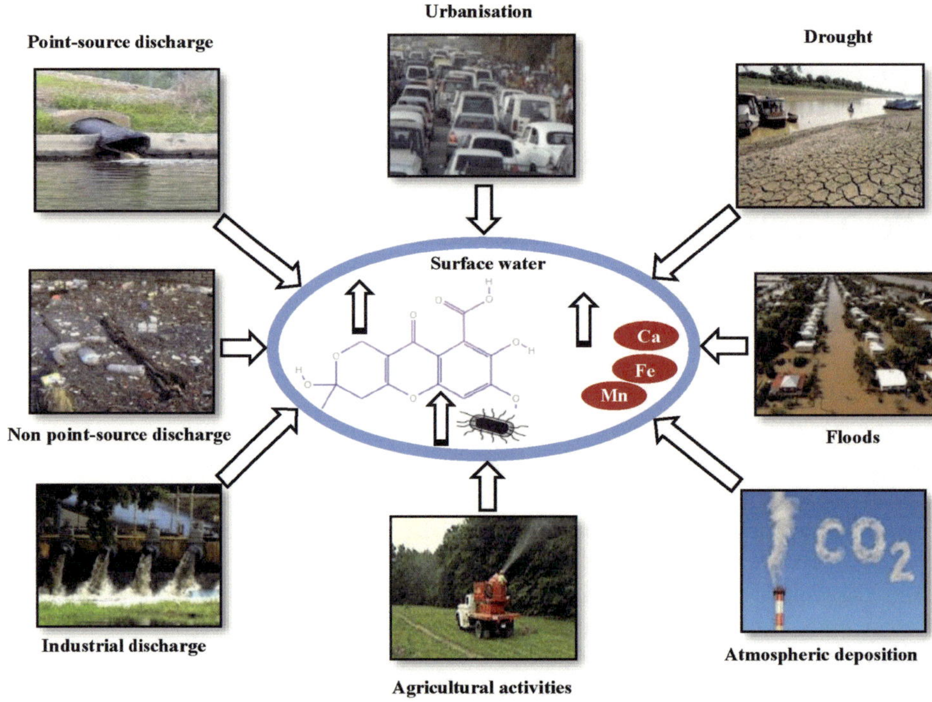

Figure 1.
Sources of surface water contamination.

cope with the increasing levels of contamination, an ultimate goal for the current research activities.

Technically, the performance of surface water treatment systems depends on the efficiency of individual treatment processes in removing contaminants. Conventional surface water treatment systems consist of coagulation/flocculation, filtration, and disinfection [1]. A number of operational and health problems arise in the surface water treatment process as a result of increasing contamination. The most common problems are high level of dangerous residual metal coagulants such as aluminum (Al) [7], fouling of filtration media [8], and the formation of hazardous disinfection by-products (DBPs) [9].

Residual metals can cause operational and health problems. Increasing the Al concentration in water increases turbidity, causes filtration fouling, and interferes with disinfectants [10–12]. In addition to the technical problems, the residual Al in treated water can cause neuropathologic disorders, neurological diseases (e.g., Alzheimer's and presenile dementia), and kidney diseases [10, 13].

Fouling of filtration/adsorption media is another challenge that is commonly encountered in potable water treatment processes. Fouling can occur as a result of the deposition of various foulants, such as solid particles, organic contaminants, inorganic contaminants, and microorganisms, onto various filter surfaces [14]. Fouling of filters results in extra cost and delay on the filtration process as well as reducing the quality of the water produced [15]. The deeply embedded microorganisms in filtration media do not only act as a hidden source of pathogens but also release toxic metabolic products into water treatment systems [16].

The formation of DBPs is a result disinfectants (e.g., chlorine and ozone) reaction with the organic matter [17, 18]. DBPs include a wide spectrum of carcinogenic and mutagenic chemical complexes that pose a threat to both humans and the environment. The two most prevalent classes of DBPs in drinking water are trihalomethanes

(THMs) and haloacetic acids (HAAs) [19]. Total THMs (TTHMs) is the sum of four compounds: chloroform, bromodichloromethane, dibromochloromethane, and bromoform [20]. HAAs include nine compounds which encompass derivatives of HAAs (i.e., mono-, di-, and trihaloacetic acid) and iodine and bromine containing HAAs [19]. The most common HAAs are di- and trihaloacetic acid. Epidemiological and toxicological studies indicated that the human exposure to chlorinated water containing DBPs may lead to bladder cancer [21], deterioration in liver functionalities, kidney and nervous system [22], and congenital diseases [17]. Therefore, a maximum contamination level (MCL) of DBPs has been set for different countries around the world. For instance, the MCL of THMs in Australia is 250 μg L^{-1}, while the MCL of monochloroacetic acid (MCAA), dichloroacetic acid (DCAA), and trichloroacetic acid (TCAA) are 150, 100, and 100 μg L^{-1}, respectively [22, 23].

2. Physical methods for drinking water treatment

Research efforts have been directed toward minimizing the challenges encountered in surface water treatment systems. It is obvious that the increasing levels of contamination and the conventional chemicals used for treatment are the main reasons behind these challenges. Hence, the quantities of chemicals added to water should be minimized without compromising the quality of the treated water. To this end, chemical-free (henceforth referred to as physical) treatment methods are recommended to be applied in surface water treatment schemes. It should be mentioned here that this study focuses on organic and microbial contamination; hence, the discussion in the following sections will be confined to aspects pertaining to the removal of such contaminants.

The common physical treatment methods include pulsed-electric field and plasma discharge [24, 25], magnetic field [26], hydrodynamic cavitation [27], ultraviolet (UV) light [28], and ultrasound [29]. The combinations of physical-physical treatments such as UV light and ultrasound and physical-chemical treatments such as ultrasound and chlorine dioxide, ultrasound and ozone, and UV and ozone are also recommended [30].

2.1 Organic contamination

The organic contamination of natural surface water is represented by the existence of natural organic matter (NOM) in water sources. NOM can be categorized based on size into particulate organic carbon (POC) and dissolved organic carbon (DOC). NOM fraction that passes through 0.45 μm filter is termed as DOC, while the retained fraction is termed as POC. The latter only forms 10% of NOM and can easily be removed from water [31]. Therefore, attention should be given to improving DOC removal from natural water.

2.1.1 DOC structure

DOC encompasses a vast array of organic materials that varies in their characteristics spatiotemporally [32]. DOC can be classified into groups based on origin and structure. Origin-based classification categorizes DOC into three groups: allochthonous, autochthonous, and anthropogenic [33]. Allochthonous is derived from natural decomposition of soil and plants, while autochthonous DOC is originated from algal and microbial activities. The anthropogenic DOC in surface water is emanated from human activities and wastewater treatment processes [33]. Potable water sources contain mainly allochthonous and autochthonous carbon [34].

The concentration of autochthonous DOC in surface water depends strongly on the hydraulic residence time of water in reservoirs and this would naturally reduce its contribution to overall organic contamination. Hence, improving allochthonous DOC removal would be of more importance to drinking water treatment practices.

The structural classification mainly divides DOC into hydrophobic and hydrophilic fractions [35]. The proportion of these fractions in natural water catchments depends on the carbon source and other factors such as microbial activities and natural photo-degradation. The hydrophobic fraction is comprised mainly of humic and fulvic acids, phenolic DOC, and double bond structures [36]. The hydrophilic fraction mainly contains aliphatic and nitrogenous compounds [35]. DOC structure is important from water treatment perspective as these fractions are associated with certain health and operational problems [37]. For instance, hydrophobic DOC is known to have a tendency to react with chlorine forming DBPs [38].

2.1.2 DOC removal mechanisms

The main DOC removal mechanisms of physical treatments are (i) chemical reactions (e.g., radicals attack), (ii) physical effects (e.g., shear forces, pyrolysis), and (iii) alteration of physical properties (absorptivity). A wide range of radicals are produced when exposing water to physical treatments such UV and ultrasound. The most important radical species is the hydroxyl ($^{\bullet}OH$) as it possesses a high oxidation potential (2.8 V) that exceeds the oxidation potentials of common oxidants such as atomic oxygen (2.42 V), ozone (2.07 V), and hydrogen peroxide (1.78 V) [39]. The $^{\bullet}OH$ pathway reactions with NOM include addition to double bonds and hydrogen and electron abstraction [35]. Chemical mechanisms are prominent in electrical and UV techniques, while the combination of both chemical and physical mechanisms is generated with techniques such as ultrasound and hydrodynamic cavitation [40]. Physical treatments that utilize magnetic fields can alter physical properties of DOC, making it more susceptible to removal via adsorption [6, 26]. It is worth mentioning that physical methods that produce $^{\bullet}OH$ are also capable of altering the nature of remnant DOC [41].

2.1.3 DOC removal with physical methods

Generally, DOC removal levels are low with the physical treatments as stand-alone technologies; however, combining these methods with chemicals addition can significantly boost DOC removal [35]. Chemical addition to some treatment methods such as UV and electrical methods can be problematic. For instance, the addition of TiO_2 in photo-catalysis (UV/semi-conductors) requires an additional treatment to remove TiO_2 particles from the treated water, and this in turn introduces extra cost [42]. The addition of electrolytes such as NaCl [35], or KCl [43] in electrochemical oxidation can also cause some technical problems such as the conformational change of DOC [44] resulting in a compact fouling layer. Electrodes and UV lamps are also prone to fouling problems that require frequent maintenance [45]. Furthermore, the use of UV method, particularly vacuum UV (VUV), was found to produce undesired nitrite by-products [35]. Similarly, magnetic field technique can potentially cause some health problems. It was reported that the use of magnetically treated water negatively affects the functionality of rats' kidneys suggesting that magnetic treatment can cause unstable changes to bio-mechanisms of tissue fluid [46]. Generally, electrical, magnetic, and UV treatments require mixing to ensure uniform effective treatments which adds to energy requirements of these techniques. By way of contrast, mixing is not required for dynamic treatments such as ultrasound and hydrodynamic cavitation. These treatments were also found to have benign environmental effects [47]. However, hydrodynamic cavitation has some disadvantages such as the unclear effect of operating parameters on cavitation events [48], the requirement of long treatment time to achieve perceptible change, and mechanical erosion of equipment [47]. The main disadvantage of ultrasound is high operational energy demand [49], nevertheless the installation and maintenance cost is low due to its simple configuration [50]. Recent studies have reported that ultrasound is more energy efficient compared to hydrodynamic cavitation and UV in removing organic materials [25].

2.2 Microbial contamination

Various species of microbes are present in surface water. However, microbial contamination of water is normally evaluated through indicators such as total coliform and *E. coli* [51]. The mechanisms of microbial removal/inactivation using physical treatment methods are similar to those of NOM removal. The produced highly oxidative agents attach the structure of microbes weakening their resistance to the surrounding environmental conditions. Similar microbial structural damage can be induced by the strong mechanical effects such as powerful turbulences and shockwaves. Generally, UV and electrical disinfection techniques rely on chemical effects; with ultrasound and hydrodynamic cavitation, the mechanical effects have a more prominent role as opposed to thermal and chemical effects [52].

As far as the performance is concerned, UV and electrical techniques have the disadvantages of producing mutagenic activities and low performance with turbid water [14, 53]. Hydrodynamic cavitation has some shortcomings as mentioned in Section 2.1.3. In contrast, ultrasound technology has advantages of being environmentally friendly and easy to implement and control, which outweighs the disadvantage of high energy demand. Even the high energy demand reputation for ultrasound technology may be attributed to the inefficient utilization of energy in this technology which will be discussed further in the coming sections.

Given the potential of ultrasound technology in solving the emerging problems in drinking water treatment process, this chapter will provide critical review on this matter.

3. Ultrasound technology

3.1 Fundamentals of ultrasound

Ultrasound is a longitudinal wave with frequency ranges between 16 kHz and 500 MHz [54]. The propagation of ultrasound waves through water produces alternating cycles of positive and negative pressure. When the magnitude of the ultrasonic pressure exceeds the tensile strength of the liquid, cavitational bubbles are created. The formed cavitational bubbles and existing gas bubbles in the liquid grow to a size larger than their original size during the negative cycle of the ultrasonic pressure. Some bubbles grow to a very large size due to gas transfer across bubble skin (rectified diffusion) or coalescence with other bubbles, and eventually float to water surface. Other bubbles collapse during the positive cycle of the ultrasonic wave. In terms of collapse intensity, there are two kinds of bubbles; bubbles with gentle collapse "stable bubbles" and bubbles with severe collapse "transit bubbles" [55]. There are two sources for bubbles generated in ultrasonically excited water: dissolved gas and gas entrapped in crevices of solid surfaces. The formation of bubbles from dissolved gas is normally termed as homogeneous cavitation, while bubbles formation on liquid-solid interface is termed as heterogeneous cavitation [56].

The physics and chemistry of transit bubbles are of interest from water treatment perspective owing to the powerful effects produced from such bubbles collapse. These effects are represented by the generation of localized areas of high temperature and pressure of around 5000 K and 500 atm, respectively, usually referred to as hot spots [40]. There is a variation in the temperature profile within the localized areas of hot spots which determines the nature of reactions occurring in each area. The three recognized zones of the hot spots are [40, 57]:

A. **Thermolytic center** represents the center of the cavitational bubble. During bubble collapse, the temperature and the pressure of this zone reach approximately 5000 K and 500 atm, respectively. The materials phase in this region is gaseous, so it can be inferred that the high temperatures in this region can lead to the thermolysis of the volatile DOC and water vapor exist in the region [58]. The thermolysis of water vapor produces free radicals that can further decompose volatile DOC.

B. **Interfacial zone** is present between bubble skin and the bulk solution. The thickness of this region is around 200 nm, and the life time of this region is about 2 μs [57]. The temperature in this region reaches to approximately 2000 K at the final collapse of the bubble [59]. The material phase in this region is a supercritical fluid. The high temperature in the interfacial zone facilitates the thermolysis and the oxidation of nonvolatile DOC.

C. **Bulk solution region**: the pressure in this region is equal to the ambient pressure; whereas, the temperature is variable depending on ultrasound operating parameters. The hydroxyl radicals recombine in the bulk solution region producing hydrogen peroxide, which in turn can oxidize nonvolatile DOC.

Bubble's oscillation and collapse generate acoustic streaming, microstreaming, microjetting, turbulence, shock wave, and shear stress [60]. Acoustic streaming is defined as the convective liquid motion due to the passage of ultrasound waves. Microstreaming is the liquid motion in the adjacent area to oscillating bubbles. Microjetting is the resulting liquid motion from bubble symmetrical collapse close to the solid/liquid interface [61]. The physical and chemical effects of ultrasound can be harnessed for organic and microbial contamination removal.

3.2 Effects of acoustic cavitation events on water contaminants

Figure 2 illustrates the physical and chemical effects of ultrasound on water contaminants. The physical effects such as the powerful turbulences and shock waves can disintegrate organic and microbial structures, as reported by several studies [49, 60].

Chemical effects of ultrasound are evident through the liberation of highly reactive species that have the capacity to cleave chemical bonds. The reactive species are short lived intermediates [62]; therefore, their effect is expected to occur only during the short time of the bubble's collapse. As explained earlier, volatile compounds are likely to decompose in the thermolytic center due to the effects of free radicals.

The nonvolatile compounds in water are divided into two groups: hydrophobic and hydrophilic compounds. The repulsive nature of hydrophobic compounds to water forces these compounds to accumulate in the area adjacent to collapsing bubbles, which in turn facilitates the ultrasonic-induced chemical decomposition of these compounds by free radicals, as demonstrated in **Figure 2**. The case is different for nonvolatile hydrophilic compounds, as the concentration of such compounds in the sheath around the bubble is similar to that in the bulk solution region. So the hydrophilic compounds are either chemically disintegrated by free radicals and their

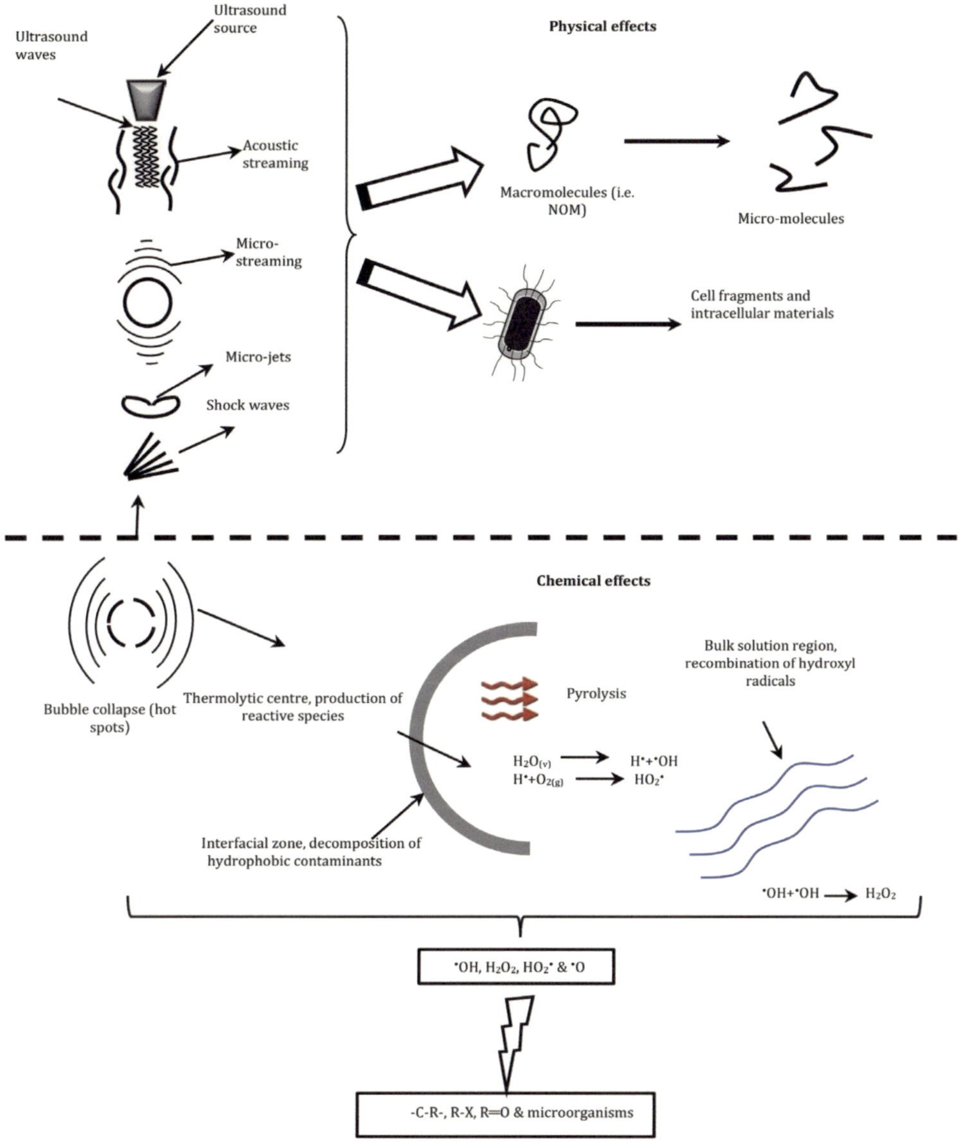

Figure 2.
Mechanisms of acoustic cavitation in degrading water pollutants.

recombination products or mechanically destructed via the mechanical shear and shock waves resulting from bubble oscillations and collapse [63]. The shear stresses' and shock waves' degradation of organic materials is attributed to the slight phase difference, especially for humic polymeric structures. Many researchers have reported the capacity of shear stresses and shockwaves on breaking the chain structure of polymeric organic materials or opening the ring structure of cyclic organic materials [57]. Additionally, the extreme conditions in the collapsing bubble's center and the surrounding areas can lead to the formation of acids [64], which can reduce the solubility of humic acid and consequently increases its degradation by the physical effects.

Although inorganic contaminants are outside the scope of this study, it is worth mentioning that microstreaming and generated oxidative species instigated by bubble collapse are the main ultrasonic removal mechanisms for these contaminants [65].

3.3 Methods of producing ultrasound waves

Ultrasound waves are commonly generated by converting electrical power into vibration using transducers. There are two types of transducers: piezoelectric and magnetostrictive [66]. A graphical representation of these transducers is shown in **Figure 3**. For piezoelectric transducers, the vibration is created via exciting the piezoelectric crystal with electrical current, as demonstrated in **Figure 3a**. In the case of magnetostrictive transducers, the electrical current is passed through coils inducing a magnetic field that causes contraction and expansion of the ferromagnetic core (Terfenol-D of Nickel in most cases), as shown in **Figure 3b**. Comprehensive comparison between the characteristics of magnetostrictive and piezoelectric transducers is provided in [67]. Although the performance of magnetostrictive transducers outstrips that of piezoelectric transducers [68, 69], there is limited number of studies concerning the use of these transducers for water treatment applications.

3.4 Modes of operation

Ultrasound irradiation can be applied in two modes: continuous and pulsed. Continuous mode is more commonly used for water treatment application compared to the pulsed mode. In pulsed mode, the operation is interrupted for a preset amount of time. The period during which ultrasound operates is known as pulse; whereas, the interruption time is normally termed as interval. The pulse and interval are denoted, respectively, as *On* and *Off* periods. The *On:Off* ratio is commonly denoted as *R*. Operating ultrasound in a pulsed mode is more energy-efficient due

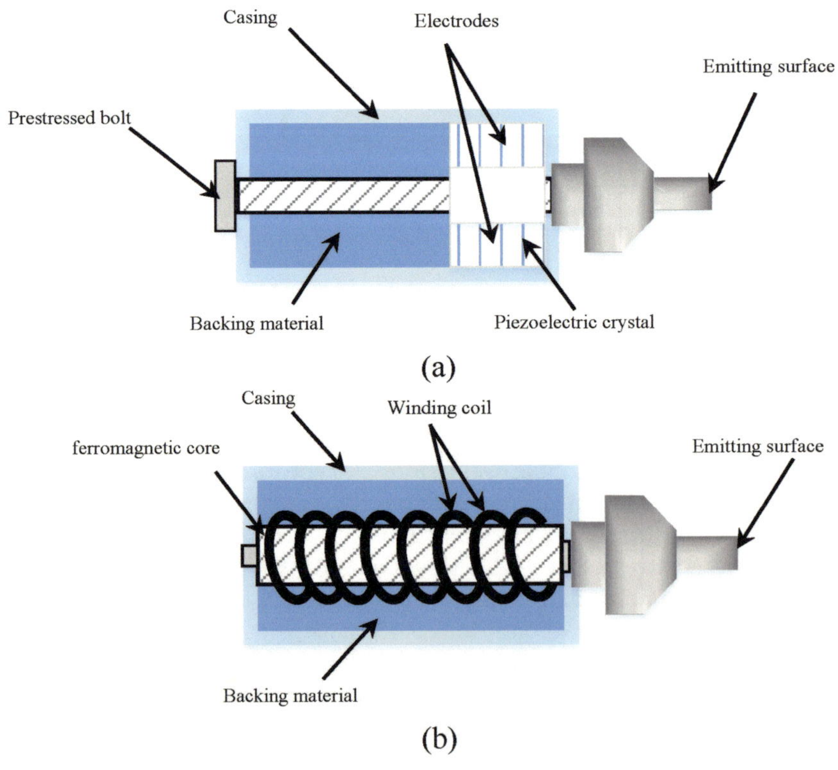

Figure 3.
Common ultrasonic wave generation techniques: (a) piezoelectric and (b) magnetostrictive.

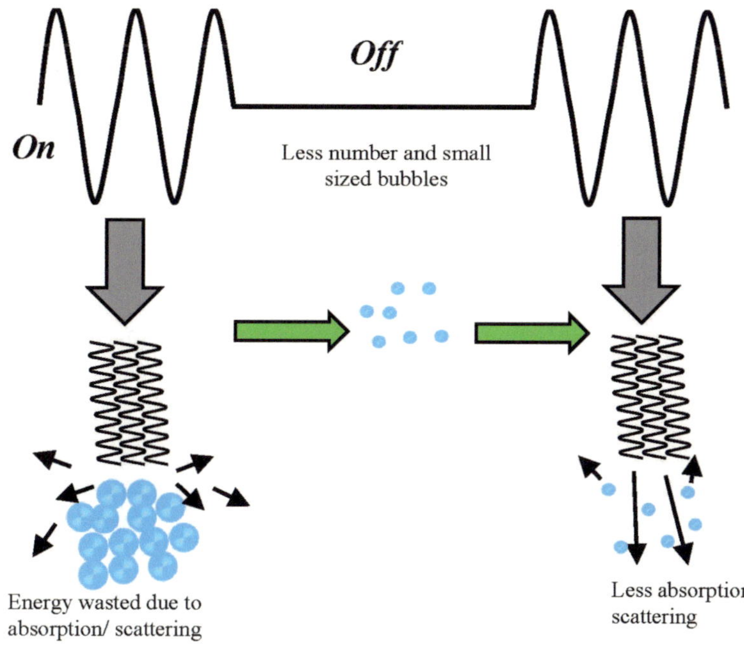

Figure 4.
Illustration of pulsed mode alleviation of shielding effects.

to minimizing bubble's cloud size that occurs near the irradiating surface especially at high-power levels (reduction of shielding effects) [57]. During the *Off* period, the ineffective cloud bubbles dissolve and/or float to the surface leaving less number of ineffective bubbles close to the irradiating surface, which means less energy is absorbed/scattered by bubbles [70], as illustrated in **Figure 4**. Other positive aspects of applying pulsed mode ultrasound include improvement of pollutants transport to reaction sites of collapsing bubbles, spatial enlargement of the active zone, and utilization of acoustic residual energy during the *Off* period. Operating ultrasound in pulsed mode also reduces temperature rise that can be undesirable for some water treatment applications such as filtration [14].

Operating ultrasound in pulsed mode does not always result in improved performance [71]; it depends on applying a suitable power level for the chosen R ratio. Hence, optimizing pulse ratios and power levels are of utmost importance for pulsed ultrasound applications. Using pulsed ultrasound for water contaminants removal was investigated by a limited number of studies, such as the studies conducted by [72, 73]. These studies dealt only with synthetic water samples. Recent studies proved the capability of pulsed ultrasound in removing natural water contaminants [74].

3.5 Parameters affecting ultrasound effectiveness

Like other treatment technologies, the performance of ultrasound is influenced by several factors. These factors can be broken down into three groups: system operating conditions, medium characteristics, and design-related aspects. The operating parameters of ultrasonic equipment include power, frequency, treatment time, mode of operation, and shape of the exciting waves (i.e., sine, triangle, etc.). It is known that increasing the power results in more intense ultrasonic effects; however, power impact normally follows a logarithmic growth trend, where increasing beyond a certain limit can only results in little improvement. Frequency has a direct relationship with cavitation threshold; therefore, the higher the frequency, the more the power required to generate cavitation bubbles [75]. As discussed in the previous section, pulsed mode is more energy-efficient than the continuous mode. Among the common exciting waves' shapes, square wave has the highest ultrasonic effects [67].

Medium characteristics such as viscosity, pressure, temperature, and contents of solid and gas impurities can affect the intensity of ultrasound effects. Viscosity has a negative effect on the generation and collapse of cavitating bubbles. It is difficult for ultrasonic waves to propagate through a viscous medium due to high cohesion forces; hence, less effective acoustic events would be achieved [76]. In the case of typical surface water treatment system, change in water viscosity is not expected to occur, and hence the effect of this factor can be ignored. The effect of the ambient pressure on ultrasound comes into play only when dealing with closed system treatment chambers. Increasing the ambient pressure has two conflicting effects: decreases the vapor content in the collapsing bubble leading to more effective bubble collapse [54] and at the same time negatively affects bubble growth leading to less violent collapse [77]. The ambient temperature impacts ultrasound performance in a similar fashion. Increasing the temperature facilitates bubbles formation due to reduction in medium viscosity; however, the vapor content in the formed bubbles would be high leading to a less violent collapse (cushioning effects) [77]. It should be mentioned that increasing the ambient temperature can accelerate both microbial disruption and chemical reactions under the effect of ultrasound [54, 77]. This means that the net temperature effect on ultrasound performance is positive.

The impact of solid particles and dissolved gas bubbles depends on their nature and the treatment purpose. Bubbles formed from gases with high specific heat ratio produce better cavitation effects (higher temperature and larger number of radicals) compared to those generated from gases with low specific heat ratio [78]. The presence of solid particles in water can be beneficial if the treatment is targeting microbes' removal [79, 80], or adverse if the treatment goal is DOC removal [81]. In the case of surface water treatment, the dissolved gas would mostly be air resulting in relatively high acoustic effects compared to other gases such as O_2 and Ar [82]. The presence of solids in surface water is inevitable, and they would be a mixture of soil aggregates that release DOC upon ultrasound exposure [81] and solid particles that promote heterogeneous cavitation [80].

The aspects of ultrasonic reactor design such as reactor shape and liquid height play crucial roles in the homogeneity of acoustic energy distribution and the uniformity of treatment across the treated volume. Generally, reactors with curvatures (e.g., conical or cylindrical) are more effective in utilizing ultrasound power compared to the standard rectangular-shaped reactors [83, 84]. This is attributed to the reflection of the waves back from the curved walls to the water in different directions resulting in more acoustic events. However, reactors with flat surfaces are easier to design and modify to accommodate monitoring and measurements equipment [57]. An example of such a design is the hexagonal reactor proposed by Gogate et al. [85], where waves can still be reflected from the walls. The liquid height has a negative effect on ultrasound performance; the further away the contaminants are from ultrasonic source, the less effective the treatment is [57]. Interestingly though, in a study conducted by Asakura et al. [86] on the effect of liquid height on ultrasound chemical activity at different frequencies showed that at largest height investigated (500 mm), low frequency ultrasound resulted in the highest chemical throughput compared to other tested frequencies (>100 kHz). In the same manner, Sharma and Sanghi [87] reported that low frequency results in better distribution of acoustic energy in large-scale volumes. This suggests that low frequency ultrasound operation has the potential to be successfully scaled up to industrial levels.

3.6 Ultrasound scalability in surface water treatment

The scalability of ultrasound technology for drinking water treatment purposes requires multi-disciplinary expertise such as chemistry, electrical engineering, chemical engineering, material sciences, etc. One essential step toward scalability is applying an accurate energy characterization technique. The use of an inappropriate characterization method would produce discouraging energy figures that would be disincentive for industries interested in adopting ultrasound technology.

There are many techniques for determining the capacity of ultrasound equipment in converting electrical power to useful acoustic energy. Among all the reported energy characterization techniques, calorimetric technique is the most commonly used owing to its simplicity and cost-effectiveness [88]. However, this technique must be carefully applied. The use of a single location for temperature measurements as being representative for the whole irradiated volume is not appropriate, especially for low power levels where standing wave effects are evident [89]. The other aspect that needs to be carefully considered is the heat loss via convection during the time of temperature recording. Convective heat loss would be more noticeable in the cases of high-power application and pulsed operation. At high ultrasonic power, the temperature rise is rapid which would accelerate thermal energy dissipation through the walls of the containing vessel to the atmosphere. In the case of pulsed ultrasound, long irradiation time is required to obtain tangible temperature rise and this would allow enough time for the generated heat to escape to the atmosphere. This explains why some studies have reported efficiency as low as 30% for ultrasonic horn [90], while others reported efficiency as high as 60–70% [91] for the same reactor type, as the latter used a sophisticated adiabatic reaction vessel that prevents convective heat loss.

Many scale-up attempts of ultrasonic reactors were reported in the literature [92]. The prominent approaches were: multistage reactors [49], flow-cells [93], sonitube [89], super-positioning multiple transducers of similar or different frequencies [57], and the use of reflectors [94, 95]. The approach of combined multi-transducers and reflectors seems to be a promising strategy for ultrasonic reactor scale-up as the interaction of waves emitted from transducers and the reflected waves from reflectors would enlarge the active zone in the reactor. However, it is worth mentioning that most of these scale-up attempts utilized the commercially available piezoelectric transducers that operate largely on sine wave excitation. Recent studies have shown that some waveforms other than the sine wave can result in better excitation of transducers [96]. Thus, exploring the use of other transducer types and waveforms in large-scale applications is imperative to provide broader and may be more efficient options to industry.

3.7 Ultrasound application in water treatment processes

3.7.1 Coagulation/flocculation

The common use of ultrasound in coagulation process is as a pre-treatment for the process to improve blue-green algae removal [97]. The presence of blue-green algae in the water treatment system has been associated with many problems such as clogging membrane pores, undesirable taste and odor, production of DBPs, and the release of toxic compounds such as Microcystin [98]. Ultrasonic mechanism for algae removal is ascribed to the destruction of gas vacuoles that are responsible of algae buoyancy [97]. There is also a recent study that has utilized ultrasound as a mean of mixing for algae removal using chitosan [99]. Removing algae requires applying low frequency, moderate input power, and short treatment time.

The application of low power ultrasound for a short treatment time in algae removal applications can solve the seasonal problem of algal bloom, but it does not tackle the problems of other forms of contamination that occur all year around. For better implementation of ultrasound in water treatment, the use of moderate to high ultrasonic power and long treatment should be applied for such applications. There is a very limited work conducted on the use of high-power ultrasound in combination with coagulation such as the work performed by Ziylan and Ince [100]. However, this work only focused on DOC removal levels, while DOC structural change and downstream effects of the treatment were not investigated. These factors were explored in [74], and it was found that ultrasound is not only capable of removing contaminants, but it also alters the structure of remnant contaminants making them more amenable to downstream treatment processes. It was also observed that ultrasound application eliminated scum formation and resulted in more compact coagulation/flocculation sludge.

3.7.2 Filtration

Ultrasound technology has been harnessed by many investigations for alleviating fouling problems in membrane filtration. Ultrasound-assisted membrane technology can be applied in two ways: cleaning or pre-treatment techniques. Ultrasonic cleaning of membrane filtration can be performed directly or indirectly. In direct ultrasonic-membrane cleaning, there is no barrier that isolates the membrane from ultrasound irradiation [57]. In an indirect ultrasonic-membrane cleaning, the membrane is isolated from ultrasonic irradiation by the membrane cell body. Most of the reports regarding ultrasound-cleaning membranes dealt with flat sheet membranes; however, in a few cases, ultrasound was also used for cleaning hollow fiber membrane modules [101] and capillary membrane fibers [102].

Although ultrasonic cleaning has been recognized by many studies as an effective alternative to chemical cleaning, there are still some shortcomings that limit its application in membrane fouling control such as dependence of cleaning effectiveness on the distance between the effective cavitational region and membrane and the detrimental effect on membrane construction materials, as shown in **Figure 5**. Deteriorating the

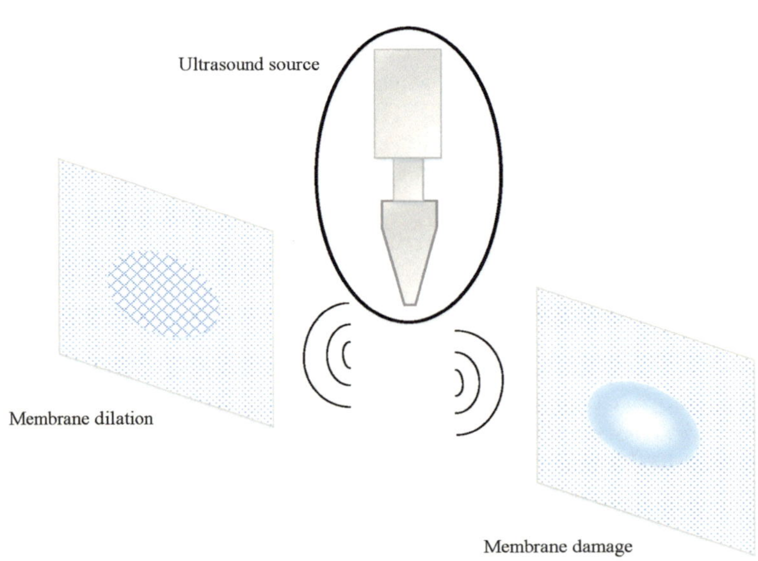

Figure 5.
Illustration of negative effects of direct high-power ultrasound on membrane structure.

structure of the membrane filter could potentially lead to a failure in filtration. Thus, the direct interaction between ultrasonic irradiation and membrane should be avoided, especially for high-power applications (up to and beyond cavitation).

As a pre-filtration process, it was found that ultrasound is capable of reducing bio-fouling formation in membrane systems [103]. Ultrasound can also remove other contaminants, as indicated in **Figure 2**. In spite of the advantages of ultrasound as a filtration pre-treatment, there are some concerns related to the disintegration of the contaminants into smaller sizes, which may then lead to a pore-plugging type of fouling [104]. For this reason, distancing ultrasound from the filtration process is recommended.

3.7.3 Disinfection

Ultrasound is recognized as the most effective disinfection technique for all forms of microbial contamination even for recalcitrant microbes and spores [47, 49, 77, 105–107]. As explained in Section 3.2, the powerful biocidal effects of ultrasound are attributed to the strong chemical and mechanical effects produced from cavitational bubble's collapse. Disinfection is typically applied after filtration at the end of the surface water treatment process. The purpose of disinfection is to disinfect water onsite and prevent microbial growth in the water while moving within the distribution network. However, as ultrasound has no residual effect, it would be more beneficial to apply ultrasound in the earlier stages of surface water treatment.

4. Conclusions and recommendations

The recent challenges in drinking water treatment industry emanating from the ever-increasing contamination sources and the application of traditional chemical treatment methods have been highlighted in this chapter. Integrating physical techniques into the conventional drinking water treatment scheme has been proposed as a potential solution for these challenges. Among the common physical techniques, ultrasound technology appears to be the most promising option. Ultrasound can produce powerful effects associated with the generation and collapse of unstable bubbles. These effects are capable of destructing microbes and mineralize organic contaminants through the production of highly oxidant species and strong mechanical effects. Appropriate utilization of ultrasound effects can only be achieved through understanding the relationship between ultrasonic parameters and the properties of the water being treated. The effect of some ultrasonic parameters such as power and frequency are extensively investigated for different treatment goals; however, this chapter attempts to draw the attention to other equally important parameters such as techniques of ultrasonic wave generation, mode of operation, and the shape of the generated waves. It appears that the best ultrasonic settings for water treatment application are moderate to high power for long treatment time, low frequency, pulsed mode, and square wave generated using magnetostrictive transducer. After critical evaluation of the possible combination scenarios of ultrasound with main drinking water treatment processes, it was concluded that applying ultrasound prior to coagulation is the most beneficial option as other combinations may create adverse downstream effects. Hence, further in-depth investigation for the suggested combination is recommended for future research work.

Conflict of interest

The authors declare no conflict of interest.

Author details

Raed A. Al-Juboori* and Les Bowtell
Faculty of Health, Engineering and Sciences, University of Southern Queensland, Toowoomba, QLD, Australia

*Address all correspondence to: raedahmed.mahmood@gmail.com

References

[1] Tansel B. New technologies for water and wastewater treatment: A survey of recent patents. Recent Patents on Chemical Engineering. 2008;**1**:17-26

[2] Wu P-H, Cheng Y-C, Chen H-Y, Chueh T-w, Chen H-C, Huang L-H, et al. Using the entrapped bioprocess as the pretreatment method for the drinking water treatment receiving eutrophic source water. Environmental Pollution. 2019;**248**:57-65

[3] Tetzlaff D, Carey S, Soulsby C. Catchments in the future North: Interdisciplinary science for sustainable management in the 21st century. Hydrological Processes. 2013;**27**:635-639

[4] Kicklighter DW, Hayes DJ, McClelland J, Peterson BJ, McGuire AD, Melillo JM. Insights and issues with simulating terrestrial DOC loading of arctic river networks. Ecological Applications. 2013;**23**:1817-1836

[5] Delpla I, Jung AV, Baures E, Clement M, Thomas O. Impacts of climate change on surface water quality in relation to drinking water production. Environment International. 2009;**35**:1225-1233

[6] Blanco J, Malato S, Fernández-Ibañez P, Alarcón D, Gernjak W, Maldonado MI. Review of feasible solar energy applications to water processes. Renewable and Sustainable Energy Reviews. 2009;**13**:1437-1445

[7] Srinivasan P, Viraraghavan T, Subramanian K. Aluminium in drinking water: An overview. Water SA. 1999;**25**:47-55

[8] Zularisam AW, Ismail AF, Salim R. Behaviours of natural organic matter in membrane filtration for surface water treatment—A review. Desalination. 2006;**194**:211-231

[9] Kim J, Chung Y, Shin D, Kim M, Lee Y, Lim Y, et al. Chlorination by-products in surface water treatment process. Desalination. 2003;**151**:1-9

[10] Driscoll C, Letterman R. Chemistry and fate of Al(III) in treated drinking water. Journal of Environmental Engineering. 1988;**114**:21-37

[11] Gabelich CJ, Yun TI, Coffey BM, Suffet IHM. Effects of aluminum sulfate and ferric chloride coagulant residuals on polyamide membrane performance. Desalination. 2002;**150**:15-30

[12] Costello JJ. Post precipitation in distribution systems. Journal of American Water Works Association. 1984;**76**:46-49

[13] Wasana HS, Perera GRK, De Gunawardena P, Bandara J. The impact of aluminum, fluoride, and aluminum–fluoride complexes in drinking water on chronic kidney disease. Environmental Science and Pollution Research. 2015;**22**:11001-11009

[14] Al-Juboori RA, Yusaf T. Biofouling in RO system: Mechanisms, monitoring and controlling. Desalination. 2012;**302**:1-23

[15] Yang T, Wan CF, Xiong JY, Chung T-S. Pre-treatment of wastewater retentate to mitigate fouling on the pressure retarded osmosis (PRO) process. Separation and Purification Technology. 2019;**215**:390-397

[16] Wingender J. In: Flemming H-C, Wingender J, Szewzyk U, editors. Hygienically Relevant Microorganisms in Biofilms of Man-Made Water Systems Biofilm Highlights. Berlin/Heidelberg: Springer; 2011. pp. 189-238

[17] Richardson SD. Disinfection by-products and other emerging contaminants in drinking water.

Trends in Analytical Chemistry. 2003;**22**:666-684

[18] Al-Jubooria RA, Yusafb T, Bowtellc LJD. Pulsed ultrasound as an energy saving mode for ultrasound treatment of surface water with terrestrial aquatic carbon. Desalination and Water Treatment. 2018;**135**:167-176

[19] Farré MJ, King H, Keller J, Gernjak W, Knight N, Watson K, et al. Disinfection by-products in South East Queensland: Assessing potential effects of transforming disinfectants in the SEQ water grid. In: Begbie DK, Wakem SL, editors. Urban Water Security Research Alliance Science Forum and Stakeholder Engagement, Biulding Linkages, Collaboration and Science Quality. Brisbane, Australia: CSIRO Publishing; 2011. pp. 94-101

[20] Clesceri LS, Rice EW, Greenberg AE, Eaton AD. Standard Methods for Examination of Water and Wastewater: Centennial Edition. Washington, D.C.: American Public Health Association; 2005

[21] Neale PA, Antony A, Bartkow ME, Farré MJ, Heitz A, Kristiana I, et al. Bioanalytical assessment of the formation of disinfection byproducts in a drinking water treatment plant. Environmental Science & Technology. 2012;**46**:10317-10325

[22] Sadiq R, Rodriguez MJ. Fuzzy synthetic evaluation of disinfection by-products—A risk-based indexing system. Journal of Environmental Management. 2004;**73**:1-13

[23] Australian National Health and Medical Research Council. Australian Drinking Water Guidelines 6. Canberra, Australia: National Resource Management Ministerial Council, Commonwealth of Australia; 2011

[24] Gusbeth C, Frey W, Volkmann H, Schwartz T, Bluhm H. Pulsed electric field treatment for bacteria reduction and its impact on hospital wastewater. Chemosphere. 2009;**75**:228-233

[25] Stratton GR, Bellona CL, Dai F, Holsen TM, Thagard SM. Plasma-based water treatment: Conception and application of a new general principle for reactor design. Chemical Engineering Journal. 2015;**273**:543-550

[26] Ambashta RD, Sillanpää M. Water purification using magnetic assistance: A review. Journal of Hazardous Materials. 2010;**180**:38-49

[27] Sawant SS, Anil AC, Krishnamurthy V, Gaonkar C, Kolwalkar J, Khandeparker L, et al. Effect of hydrodynamic cavitation on zooplankton: A tool for disinfection. Biochemical Engineering Journal. 2008;**42**:320-328

[28] Armstrong GN, Watson IA, Stewart-Tull DE. Inactivation of *B. cereus* spores on agar, stainless steel or in water with a combination of Nd:YAG laser and UV irradiation. Innovative Food Science and Emerging Technologies. 2006;7:94-99

[29] Mason TJ, Joyce E, Phull SS, Lorimer JP. Potential uses of ultrasound in the biological decontamination of water. Ultrasonics Sonochemistry. 2003;**10**:319-323

[30] Zou H, Tang HJW. Comparison of different bacteria inactivation by a novel continuous-flow ultrasound/chlorination water treatment system in a pilot scale. Water. 2019;**11**:258

[31] Leenheer JA, Croué J-P. Peer reviewed: Characterizing aquatic dissolved organic matter. Environmental Science & Technology. 2003;**37**:18A-26A

[32] Sharp EL, Parsons SA, Jefferson B. Seasonal variations in

natural organic matter and its impact on coagulation in water treatment. Science of the Total Environment. 2006;**363**:183-194

[33] Knapik H, Fernandes CS, de Azevedo J, dos Santos M, Dall'Agnol P, Fontane D. Biodegradability of anthropogenic organic matter in polluted rivers using fluorescence, UV, and BDOC measurements. Environmental Monitoring and Assessment. 2015;**187**:1-15

[34] Park HK, Byeon MS, Shin YN, Jung DI. Sources and spatial and temporal characteristics of organic carbon in two large reservoirs with contrasting hydrologic characteristics. Water Resources Research. 2009;**45**:411-418

[35] Matilainen A, Sillanpää M. Removal of natural organic matter from drinking water by advanced oxidation processes. Chemosphere. 2010;**80**:351-365

[36] Ghernaout D. The hydrophilic/hydrophobic ratio vs. dissolved organics removal by coagulation—A review. Journal of King Saud University - Science. 2014;**26**:169-180

[37] Xing L, Murshed MF, Lo T, Fabris R, Chow CWK, van Leeuwen J, et al. Characterization of organic matter in alum treated drinking water using high performance liquid chromatography and resin fractionation. Chemical Engineering Journal. 2012;**192**:186-191

[38] Soh YC, Roddick F, van Leeuwen J. The impact of alum coagulation on the character, biodegradability and disinfection by-product formation potential of reservoir natural organic matter (NOM) fractions. Water Science and Technology. 2008;**58**:1173-1179

[39] Hartnett JP, Fridman A, Cho YI, Greene GA, Bar-Cohen A. Transport Phenomena in Plasma. London, UK: Elsevier Science; 2007

[40] Lifka J, Ondruschka B, Hofmann J. The use of ultrasound for the degradation of pollutants in water: Aquasonolysis—A review. Engineering in Life Sciences. 2003;**3**:253-262

[41] Al-Juboori RA, Yusaf T, Aravinthan V, Bowtell L. Investigating natural organic carbon removal and structural alteration induced by pulsed ultrasound. Science of the Total Environment. 2016;**541**:1019-1030

[42] Klavarioti M, Mantzavinos D, Kassinos D. Removal of residual pharmaceuticals from aqueous systems by advanced oxidation processes. Environment International. 2009;**35**:402-417

[43] Motheo AJ, Pinhedo L. Electrochemical degradation of humic acid. Science of the Total Environment. 2000;**256**:67-76

[44] Ghosh K, Schnitzer M. Macromolecular structures of humic substances. Soil Science. 1980;**129**:266-276

[45] Zhe C, Hong-Wu W, Lu-Ming M. Research progress on electrochemical disinfection technology for water treatment. Industrial Water and Wastewater. 2008;**39**:1-5

[46] Singh M, Singh U, Singh K, Mishra A. Effect of 50-Hz powerline exposed magnetized water on rat kidney. Electromagnetic Biology and Medicine. 2004;**23**:241-249

[47] Gogate PR, Kabadi AM. A review of applications of cavitation in biochemical engineering/biotechnology. Biochemical Engineering Journal. 2009;**44**:60-72

[48] Arrojo S, Benito Y. A theoretical study of hydrodynamic cavitation. Ultrasonics Sonochemistry. 2008;**15**:203-211

[49] Hulsmans A, Joris K, Lambert N, Rediers H, Declerck P, Delaedt Y, et al. Evaluation of process parameters of ultrasonic treatment of bacterial suspensions in a pilot scale water disinfection system. Ultrasonics Sonochemistry. 2010;**17**:1004-1009

[50] Furuta M, Yamaguchi M, Tsukamoto T, Yim B, Stavarache CE, Hasiba K, et al. Inactivation of Escherichia coli by ultrasonic irradiation. Ultrasonics Sonochemistry. 2004;**11**:57-60

[51] Wright J, Gundry S, Conroy R. Household drinking water in developing countries: A systematic review of microbiological contamination between source and point-of-use. Tropical Medicine & International Health. 2004;**9**:106-117

[52] Yusaf T, Al-Juboori RA. Alternative methods of microorganism disruption for agricultural applications. Applied Energy. 2014;**114**:909-923

[53] Reyns KM, Diels AM, Michiels CW. Generation of bactericidal and mutagenic components by pulsed electric field treatment. International Journal of Food Microbiology. 2004;**93**:165-173

[54] Thompson LH, Doraiswamy LK. Sonochemistry: Science and engineering. Industrial and Engineering Chemistry Research. 1999;**38**:1215-1249

[55] Young FR. Cavitation. London: Imperial College Press; 1999

[56] Al-Juboori RA, Yusaf TF. Improving the performance of ultrasonic horn reactor for deactivating microorganisms in water. IOP Conference Series: Materials Science and Engineering. 2012;**36**:1-13

[57] Chen D, Sharma SK, Mudhoo A. Handbook on Applications of Ultrasound: Sonochemistry for Sustainability. Boca Raton: CRC Press; 2011

[58] Flint EB, Suslick KS. The temperature of cavitation. Science. 1991;**253**:1397-1399

[59] Riesz P, Berdahl D, Christman C. Free radical generation by ultrasound in aqueous and nonaqueous solutions. Environmental Health Perspectives. 1985;**64**:233

[60] Gogate PR. Application of cavitational reactors for water disinfection: Current status and path forward. Journal of Environmental Management. 2007;**85**:801-815

[61] Birkin PR, Silva-Martinez S. A study of the effect of ultrasound on mass transport to a microelectrode. Journal of Electroanalytical Chemistry. 1996;**416**:127-138

[62] Kondo T, Krishna CM, Riesz P. Free radical generation by ultrasound in aqueous solutions of nucleic acid bases and nucleosides: An ESR and spin-trapping study. International Journal of Radiation Biology. 1988;**53**:331-342

[63] Henglein A, Gutierrez M. Chemical effects of continuous and pulsed ultrasound: A comparative study of polymer degradation and iodide oxidation. The Journal of Physical Chemistry. 1990;**94**:5169-5172

[64] Feng R, Zhao Y, Zhu C, Mason TJ. Enhancement of ultrasonic cavitation yield by multi-frequency sonication. Ultrasonics Sonochemistry. 2002;**9**:231-236

[65] Nishida I. Precipitation of calcium carbonate by ultrasonic irradiation. Ultrasonics Sonochemistry. 2004;**11**:423-428

[66] Povey JW, Mason TJ. Ultrasound in Food Processing. Tunbridge wells, kent: Springer; 1998

[67] Al-Juboori RA, Bowtell LA, Yusaf T, Aravinthan V. Insights into the scalability of magnetostrictive ultrasound technology for water treatment applications. Ultrasonics Sonochemistry. 2016;**28**:357-366

[68] Claeyssen F, Colombani D, Tessereau A, Ducros B. Giant dynamic magnetostrain in rare earth-iron magnetostrictive materials. IEEE Transactions on Magnetics. 1991;**27**:5343-5345

[69] Claeyssen F, Lhermet N, Maillard T. Magnetostrictive actuators compared to piezoelectric actuators. In: European Workshop on Smart Structures in Engineering and Technology. Giens, France: SPIE; 2003. pp. 194-200

[70] Roy RA. Cavitation sonophysics. NATO ASI Series, Series C: Mathematical and Physical Sciences. 1999;**524**:25-38

[71] Gutierrez M, Henglein A. Chemical action of pulsed ultrasound: Observation of an unprecedented intensity effect. The Journal of Physical Chemistry. 1990;**94**:3625-3628

[72] Casadonte DJ, Flores M, Petrier C. The use of pulsed ultrasound technology to improve environmental remediation: A comparative study. Environmental Technology. 2005;**26**:1411-1418

[73] Ashokkumar M, Vu T, Grieser F, Weerawardena A, Anderson N, Pilkington N, et al. Ultrasonic treatment of cryptosporidium oocysts. Health Related Water Microbiology. 2003;**47**:173-177

[74] Al-Juboori RA, Aravinthan V, Yusaf T, Bowtell L. Assessing the application and downstream effects of pulsed mode ultrasound as a pre-treatment for alum coagulation. Ultrasonics Sonochemistry. 2016;**31**:7-19

[75] Al-Juboori Raed A, Yusaf T. Identifying the optimum process parameters for ultrasonic cellular disruption of *E. coli*. International Journal of Chemical Reactor Engineering. 2012

[76] Leadley CE, Williams A. Pulsed electric field processing, power ultrasound and other emerging technologies. In: Brennan JG, editor. Food Processing: Handbook. Weinheim, Germany: Wiley-VCH Verlag GmbH & Co. KGaA; 2006

[77] Al-Juboori RA, Yusaf T. Identifying the optimum process parameters for ultrasonic cellular disruption of *E. coli*. International Journal of Chemical Reactor Engineering. 2012;**10**:1-32

[78] Chen X. Nanoplatform-Based Molecular Imaging. New Jersey: Wiley; 2011

[79] Ince NH, Belen R. Aqueous phase disinfection with power ultrasound: Process kinetics and effect of solid catalysts. Environmental Science & Technology. 2001;**35**:1885-1888

[80] Dadjour MF, Ogino C, Matsumura S, Shimizu N. Kinetics of disinfection of *Escherichia coli* by catalytic ultrasonic irradiation with TiO_2. Biochemical Engineering Journal. 2005;**25**:243-248

[81] Bossio JP, Harry J, Kinney CA. Application of ultrasonic assisted extraction of chemically diverse organic compounds from soils and sediments. Chemosphere. 2008;**70**:858-864

[82] Entezari MH, Kruus P, Otson R. The effect of frequency on sonochemical reactions. III: Dissociation of carbon disulfide. Ultrasonics Sonochemistry. 1997;**4**:49-54

[83] Capelo JL, Galesio MM, Felisberto GM, Vaz C, Pessoa JC. Micro-focused ultrasonic solid–liquid

extraction (μFUSLE) combined with HPLC and fluorescence detection for PAHs determination in sediments: Optimization and linking with the analytical minimalism concept. Talanta. 2005;**66**:1272-1280

[84] Priego-López E, Luque de Castro MD. Ultrasound-assisted extraction of nitropolycyclic aromatic hydrocarbons from soil prior to gas chromatography-mass detection. Journal of Chromatography. A. 2003;**1018**:1-6

[85] Gogate PR, Mujumdar S, Pandit AB. Large-scale sonochemical reactors for process intensification: Design and experimental validation. Journal of Chemical Technology & Biotechnology. 2003;**78**:685-693

[86] Asakura Y, Nishida T, Matsuoka T, Koda S. Effects of ultrasonic frequency and liquid height on sonochemical efficiency of large-scale sonochemical reactors. Ultrasonics Sonochemistry. 2008;**15**:244-250

[87] Sharma SK, Sanghi R. Advances in Water Treatment and Pollution Prevention. Netherlands: Springer; 2012

[88] Taurozzi J, Hackley V, Wiesner M. Preparation of nanoparticle dispersions from powdered material using ultrasonic disruption. NIST Special Publication. 2012;**1200**:2

[89] Faïd F, Contamine F, Wilhelm AM, Delmas H. Comparison of ultrasound effects in different reactors at 20 kHz. Ultrasonics Sonochemistry. 1998;**5**:119-124

[90] Virkutyte J, Varma RS, Jegatheesan V. Treatment of Micropollutants in Water and Wastewater. London: IWA Publishing; 2010

[91] Löning J-M, Horst C, Hoffmann U. Investigations on the energy conversion in sonochemical processes. Ultrasonics Sonochemistry. 2002;**9**:169-179

[92] Gallego-Juárez JA, Graff KF. Power Ultrasonics: Applications of High-Intensity Ultrasound. Cambridge, UK: Elsevier Science; 2014

[93] Sun DW. Handbook of Frozen Food Processing and Packaging. 2nd ed. Boca Raton, FL: Taylor & Francis; 2011

[94] Seymour JD, Wallace HC, Gupta RB. Sonochemical reactions at 640 kHz using an efficient reactor. Oxidation of potassium iodide. Ultrasonics Sonochemistry. 1997;**4**:289-293

[95] Adewuyi YG, Oyenekan BA. Optimization of a sonochemical process using a novel reactor and Taguchi statistical experimental design methodology. Industrial and Engineering Chemistry Research. 2007;**46**:411-420

[96] Suomi V, Cleveland R, Edwards D. Measuring and modelling of harmonic acoustic radiation force induced deformations. In: 15th International Symposium on Therapeutic Ultrasound, International Society for Therapeutic Ultrasound (ISTU); Utrecht, Netherlands; 2015. p. 45

[97] Zhang G, Zhang P, Fan M. Ultrasound-enhanced coagulation for *Microcystis aeruginosa* removal. Ultrasonics Sonochemistry. 2009;**16**:334-338

[98] Lee TJ, Nakano K, Matsumara M. Ultrasonic irradiation for blue-green algae bloom control. Environmental Technology. 2001;**22**:383-390

[99] Fast SA, Gude VG. Ultrasound-chitosan enhanced flocculation of low algal turbid waters. Journal of Industrial and Engineering Chemistry. 2015;**24**:153-160

[100] Ziylan A, Ince NH. Ozonation-based advanced oxidation for pre-treatment of water with residuals of anti-inflammatory medication.

Chemical Engineering Journal. 2013;**220**:151-160

[101] Li X, Yu J, Nnanna AGA. Fouling mitigation for hollow-fiber UF membrane by sonication. Desalination. 2011;**281**:23-29

[102] Naddeo V, Belgiorno V, Borea L, Secondes MFN, Ballesteros F. Control of fouling formation in membrane ultrafiltration by ultrasound irradiation. Environmental Technology. 2014;**36**:1299-1307

[103] Al-Juboori RA, Yusaf T, Aravinthan V. Investigating the efficiency of thermosonication for controlling biofouling in batch membrane systems. Desalination. 2012;**286**:349-357

[104] Lim AL, Bai R. Membrane fouling and cleaning in microfiltration of activated sludge wastewater. Journal of Membrane Science. 2003;**216**:279-290

[105] Hua I, Thompson JE. Inactivation of Escherichia coli by sonication at discrete ultrasonic frequencies. Water Research. 2000;**34**:3888-3893

[106] Piyasena P, Mohareb E, McKellar RC. Inactivation of microbes using ultrasound: A review. International Journal of Food Microbiology. 2003;**87**:207-216

[107] Koda S, Miyamoto M, Toma M, Matsuoka T, Maebayashi M. Inactivation of *Escherichia coli* and *Streptococcus mutans* by ultrasound at 500 kHz. Ultrasonics Sonochemistry. 2009;**16**:655-659

Chapter 7

Application of High-Power Ultrasound in the Food Industry

Leire Astráin-Redín, Salomé Ciudad-Hidalgo, Javier Raso, Santiago Condón, Guillermo Cebrián and Ignacio Álvarez

Abstract

The purpose of this chapter is to summarize potential applications of the high-power ultrasound technology (5 W/cm^2; 20–100 kHz) in the food industry. Those applications are mainly related to the improvement in mass and energy transfer in different processes when ultrasound is applied in water or through air, e.g., reduction in dehydration; thawing and freezing times and energy costs of plant-, meat-, or fish-based products; increase the extraction yields of intracellular compounds with biological activity; reduction of chemical health risks such as cadmium or acrylamide; etc. The influence of some physical parameters like temperature and pressure in cavitation intensity and the potential of this technology to even inactivate microorganisms in food products and surfaces in contact with food will be discussed. Several examples of these applications will be presented, with reference to some of the industrial or pilot plant systems available in the market to be implemented in the food industry.

Keywords: mass transfer, heat transfer, cavitation, food preservation, food quality

1. Introduction

Ultrasound is considered an emerging technology in the food industry that is gaining interest due to its potential to improve several process including mass and energy transfer processes among others. It also enables to obtain safer and higher quality products than with traditional procedures. Furthermore, it should be remarked that it is also considered a safe, nonpolluting and environmentally friendly technology [1].

Ultrasonic technology consists of the application of mechanical waves with frequency over the threshold of human hearing (>16 kHz) [2]. Depending on its frequency and intensity, the ultrasonic spectrum can be further divided into low-frequency (20–100 kHz) high-power (>1 W/cm^2) ultrasound and high-frequency (>100 kHz) low-power (<1 W/cm^2) ultrasound. Low-power ultrasound is applied for noninvasive and nondestructive analyses, and it is mainly used in other areas such as medicine and cosmetics. In the food industry, this type of ultrasonic waves is basically used for process and quality control (e.g., fluid flow and container filling control, location of foreign bodies, or evaluation of the homogenization and/or emulsification efficiency). In contrast, high-power ultrasound is able to produce changes in the material or process to which they are applied, and it is used in a large variety of processes in the food industry (e.g., surface cleaning and decontamination, microbial and enzymatic inactivation, degassing, defoaming, and improvement of mass transfer, among others). Therefore, high-power ultrasound is the one of great interest in the food industry, and in this chapter, it will be discussed in more detail.

2. Effects of ultrasound in food matrices mechanism of action

Ultrasonic sound waves propagate through air, water, and solid media, generating pressure variations that cause the vibration of particles in the medium. The effects of the application of high-power ultrasound in food products are therefore dependent on the medium of propagation (liquid, solid or gas) and also on the parameters of the process such as frequency, intensity, pressure, and temperature, among others. Applying ultrasound in liquid medium is the simplest and the most common process in the food industry. Cavitation is the main phenomenon responsible of ultrasound effects when applied to a liquid. Basically, cavitation occurs when the microbubbles present in the liquid increase in size as a result of the cycles of high and low pressure generated by the ultrasonic waves until they become unstable and collapse releasing a large amount of energy (theoretically up to 5000 K and 1000 atm) [1]. As a consequence, different effects are generated. These can be divided into physical and chemical effects. Within the physical effects, microjets and microstreaming phenomena are the most relevant ones. Microjets are high-pressure water streams projected to the surface of solids that lead to the formation of pores and surface erosion, causing the release of material into the medium depending on the intensity of the jets. By contrast, microstreaming occurs in the middle of the surrounding liquid, and when its speed is high enough, it can break membrane cells, release intracellular enzymes, etc. [3]. These physical effects are more likely to occur at low frequencies (20–40 kHz) when the number of cavitation spots is low but the energy associated to them is higher. At higher frequencies (80–100 kHz), the number of spots is higher, but bubble size is smaller, so the energy released is lower and the prevalent effects are mainly chemical [4]. The primary radicals that are generated by ultrasound are H• and •OH, which can be then recombined to form other reactive species (H_2, H_2O_2) [5]. Therefore, depending on both the ultrasound intensity and, mainly, frequency, different effects, physical or chemical, are produced.

On the other hand, when an ultrasonic wave passes through a solid medium, it produces a series of alternating contractions and expansions, a phenomenon known as the "sponge effect," which facilitates the transfer of matter with the medium surrounding the solid [6]. Moreover, this mechanical stress can cause the formation of microchannels in the interior of the solid, also favoring mass transfer processes. In this case, it is unlikely that the cavitation phenomenon would occur in the liquid phase of the solid matrix [7].

Finally, although the application of high-intensity ultrasound is more complicated in gas medium, its effects on the solid/gas interface are particularly interesting, including pressure variation, oscillating flow, and microstreams [8]. The development of efficient ultrasonic systems to be applied in for gas medium is highly limited by the power loss that occurs when sound waves are propagating through air and by the mismatch between acoustic impedances of gases and solids or liquids [9]. As it will be discussed below, its main application is the improvement of food dehydration processes and defoaming.

3. Factors affecting cavitation

In the food industry, ultrasound is applied trough a liquid media in most applications, becoming cavitation the main mechanisms of action in these processes, as pointed out above. However, in order to apply ultrasound effectively to these food matrices, it is necessary to consider a group of factors influencing the cavitation phenomenon, including the characteristics of the ultrasound source (frequency, amplitude, ultrasonic supplier), characteristics of the treatment medium (solid particles, gas bubbles, viscosity), and treatment conditions (pressure and temperature) [2]. Regarding the characteristics of ultrasound source, the frequency and amplitude are the most important parameters that condition the effects of the treatment. As stated above, frequency determines the size of the bubbles and, thus, the intensity of the implosion. Amplitude is directly related to the amount of energy supplied to the system and the ultrasonic intensity [3]. At high amplitudes, the oscillation of the bubbles is higher, being the implosion more powerful and leading to further effects derived from cavitation. However, depending on the desired effects, this may not always be of interest, and therefore it is essential to optimize the treatment parameters. For example, for hydrating thawed cod fillets, the highest weight gain (18%) of fillets after 48 hours of hydration was observed when applying the 10% of the power of an ultrasound system of 35 kHz and 200 W. When ultrasound was working at the maximum amplitude of the system (100%), 12% of weight gain was observed, which was a lower value than that of the control process without using ultrasound (14%) [10]. As it will be discussed later on, both frequency and amplitude condition the ultrasonic supplier which defines the way of application of ultrasound to the product and its effects.

Besides the state of the treatment medium (solid, liquid or gas), solid particles, gas bubbles, and viscosity also influence cavitation. The presence of solid particles and gas bubbles act as nucleation points which enhance the formation of bubbles reducing the effects of cavitation. Regarding the viscosity of the medium, bubble formation is more difficult the higher the viscosity of the medium is, but the implosion is more powerful. Moreover, ultrasound has interesting effects in viscous products in order to improve energy transfer as it will be discussed below.

Finally, temperature and static pressure are key factors conditioning cavitation which are modified depending on the application. Thus, the increment of temperature reduces the viscosity of the medium and raises the vapor pressure enabling bubble formation. However, the amount of vapor inside the bubbles increases with temperature, producing the cushioning of the collapse and leading to a lower intensity of cavitation. Therefore, it is considered that there is an optimal temperature at which acoustic cavitation is maximum [11]. On the contrary, when pressure increases, cavitation is hindered, but when the implosion happens, the energy released is considerably higher. Based on these effects of temperature and pressure, two processes have been defined: manosonication (MS) and manothermosonication (MTS) which have been shown to offer new possibilities of ultrasound at temperatures near or even above 100°C as it will be commented later on.

In summary, many factors have to be considered when designing ultrasound equipment and processes in the food industry in order to secure an efficient application.

4. Basic ultrasound systems used in the food industry

Since the application of ultrasound in the food industry is very dependent on the ultrasound supplier, it is worthy to consider this point.

There are different ultrasound systems for food applications depending on the treatment medium and the desired effect. It is essential to achieve a successful fit between application system and treatment medium in order to be able to transfer the maximum amount of acoustic energy to the medium. As indicated, the application of ultrasound through liquid medium is the most used in the food industry. For this application, commercial equipment can be divided into two types: ultrasound water baths (indirect application) and probes or horns (direct application). Ultrasonic water baths are widely used due to their lower price and easy maintenance. They consist of a tank to which one or more piezoelectric transducers (40–130 kHz) are connected at the bottom or at the sides and the generated sound waves are propagated through the water or other liquid medium in which the food product is immersed (**Figure 1(a)**) [12]. The ultrasonic intensity is low (0.1–1 W/cm^2), and the treatment is less homogeneous throughout the volume due to the formation of nodes [11]. In the food industry, this type of equipment has been used for surface cleaning, degassing, enzymatic and microbial inactivation, improvement of mass transfer, etc. [13]. On the other hand, horn or probe is a direct system in which the food product is in contact to the ultrasound supplier. These equipment allow to apply higher intensities (>5 W/cm^2) than water baths, but they are more expensive. In these systems, three parts can be differenced (**Figure 1(b)**): the transducer, the amplifier of the ultrasonic signal, and the horn. The tip of the horn has to be introduced into the sonication medium, so this design is mainly used for treating liquid foods, but application in solids has also been described [14, 15]. Depending on the shape of the horn, its application will be determined and used for cell disruption, homogenization, cutting of soft products, etc. [16].

The equipment developed for the application of ultrasound through the air (called airborne) is less common due to the difficulty of its design. The type of transducer used for this application differs depending on the application: stepped plate, ribbed plate, stepped-ribbed plate, and cylindrical radiator [1, 13, 17]. The basic structure is a piezoelectric transducer in sandwich configuration and an amplifier or horn (**Figure 1(c)**). The horn is attached to a radiator which vibrates, and due to its surface, the resistance increases, and the differences in impedance between the transducer and medium are reduced. This kind of systems is very well described in the works of Gallego-Juárez et al. [1] and Charoux et al. [17], among other publications.

5. Applications of high-power ultrasound in the food industry

In recent years, numerous applications of high-power ultrasound have been developed in food processing, including product quality control, emulsification, food preservation, and improvement of mass and energy transfer processes. Some of these applications are summarized in this part of the chapter.

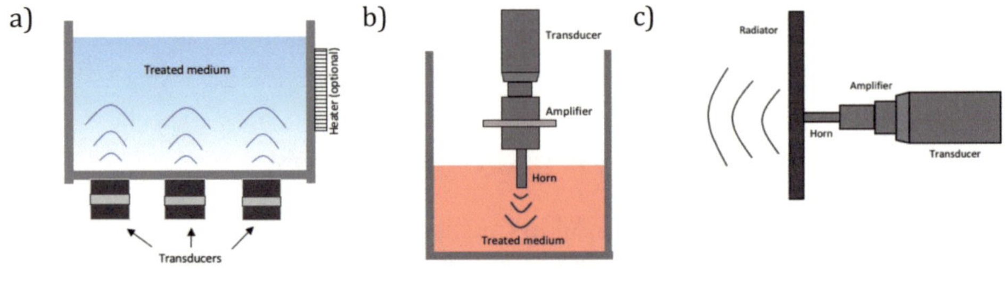

Figure 1.
Ultrasound generation systems: (a) ultrasonic bath, (b) probe or horn, and (c) airborne transducer.

5.1 Emulsion formation

The use of ultrasound for obtaining emulsions was one of the first applications in the food industry. An emulsion is a heterogeneous system formed by two immiscible liquids in which one of them is dispersed in the other in the form of small droplets with a diameter—in general—lower than 1 mm.

Li and Fogler [18, 19] originally proposed a mechanism for explaining the emulsifying capacity of ultrasound that was later confirmed by high-speed photography [20], consisting of two steps. First, the acoustic waves generate instability at the interface of the two liquids, causing large drops of oil to propel them into the aqueous phase. Second, cavitation produces microcurrents and shear forces that reduce the droplet size needed to form the emulsion [21].

There are many studies on ultrasound-assisted emulsion preparation [20, 22–24]. In general, these studies conclude that it was possible to obtain emulsions that have smaller particle size, are less polydisperse, and are more stable than by agitation by using ultrasound. For example, in a study comparing the use of ultrasound with traditional agitation [25], the application of ultrasound allowed the elaboration of a nanoemulsion of mustard oil in water with an interfacial area of 67-fold greater than that obtained mechanically. In addition, the sonicated emulsions had a narrower particle size distribution (0.82–44.6 μm) than the control emulsions (8.1–610 μm).

Due to the emulsifying capacity of ultrasounds, they are recently being used as encapsulation systems in the food industry [26]. Some high-value nutrients are encapsulated in the food matrix to avoid functional losses, organoleptic losses, undesirable reactions with other compounds, etc. Ghasemi and Abbasi [27] combined the alkalization of pH with the application of ultrasound (25 kHz, 600 W) to encapsulate oils with a high content of polyunsaturated acids in skimmed milk.

5.2 Food preservation

5.2.1 Microbial and enzyme inactivation

The main agents responsible for food spoilage are enzymes and microorganisms. Moreover, pathogenic microorganisms are responsible of food poisoning and food outbreaks, requiring therefore their control or inactivation. There are several strategies to limit their action (i.e., reducing temperature, controlling water activity, etc., of foods) and to inactivate them, mainly by heat treatments. Thermal pasteurization and sterilization are the most common technologies used for enzyme and microbial inactivation in order to obtain safe and stable food products. However, the intensity of these treatments can lead to loss of nutrients and deterioration of sensory characteristics and functional properties of food [16, 28]. Due to this, technologies which enable to inactivate those agents at lower temperatures are under evaluation being ultrasound a possibility.

Bacterial inactivation with ultrasound has been widely studied and even suggested as a possible food preservation method [29–31]. Microbial inactivation is mainly induced by the physical effects of cavitation such as shear forces, shock waves, and microcurrents that can damage cell integrity by weakening or breaking cell envelopes [32, 33]. However, its lethal effect is reduced and requires prolonged periods of time [34, 35], limiting its application as a food preservation system. Due to its low bactericidal efficacy and in order to increase its lethality, ultrasound is applied over atmospheric pressure (manosonication, MS), combined with heat (manothermosonication, MTS) and with other nonthermal technologies (pulsed electric fields, high hydrostatic pressures,

UV light) [36, 37]. From all these combinations, MS and MTS showed the most promising results since vegetative cells and even bacterial spores can be inactivated at low temperatures (40°C) [32, 33, 38], as summarized in **Figure 2**. The possibility of inactivating vegetative cells and spores opens the way to design alternative processes to thermal pasteurization and sterilization by using MTS treatments at lower temperatures than those used in traditional thermal treatments [32–33, 38]. However, the required ultrasound intensities to achieve several $\log_{10}$ cycles of microbial inactivation are still far away for its industrial application due to technical limitations.

Likewise, ultrasound is also effective for inactivating enzymes, but very long processing times are required. However, when combined with heat (thermo-sonication, TS), pressure (MS), or heat and pressure (MTS), processing times can also be reduced. For example, the application of MTS is able to reduce the heat resistance of enzymes by 2–400-fold such as alkaline phosphatase, polyphenol oxidase, peroxidase, lipase, lipoxygenase, pectin methylesterase, and polygalacturonase compared to heat treatments applied at the same temperature [32, 39–45]. As an example, **Figure 3** shows the activity reduction of pectin methylesterase of tomato juice treated by heat, MS, and MTS treatments at 62.5°C and 1 minute. As can be observed, the MTS treatment led to a complete inactivation of the enzyme, being this effect higher than the addition of the heat and MS inactivation effects when applied separately (synergistic effect).

5.2.2 Microbial decontamination and surface cleaning

Cleaning and decontamination of food equipment and/or surfaces in contact with food are among the first applications of ultrasound in the food industry besides emulsification. The main phenomena responsible for its effect are cavitation and microstreaming formed in the washing liquid. The collapse of the bubbles generates high-pressure microjets that impact the surface which favor the dissolution of compounds and the release of the particles (including microorganisms) adhered to the solid. The surfaces of the solids have irregularities and pores limiting the cleaning effectivity of traditional systems. However, ultrasounds are able to get access and get a deeper cleaning enhancing also the effectiveness of chemical cleaning by favoring the release of contaminants such as oils, proteins, and even microbial biofilms, making them more accessible to chemicals [12, 46]. Nevertheless, it is important to notice that as the ultrasonic field is not uniform throughout

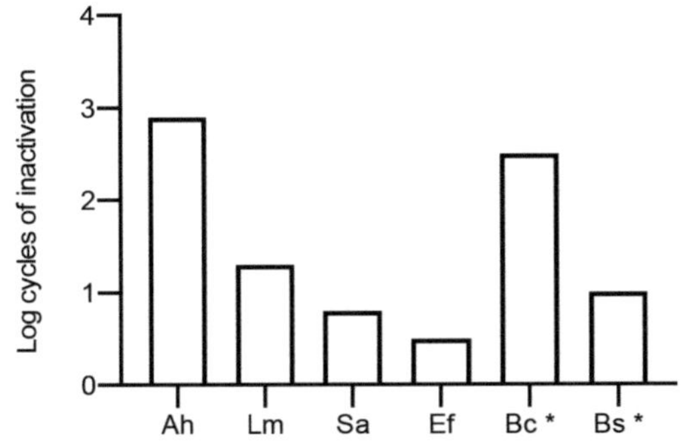

Figure 2.
Log_{10} cycles of inactivation of Aeromonas hydrophila *(ah),* Listeria monocytogenes *(lm),* Staphylococcus aureus *(Sa),* Enterococcus faecium *(Ef),* Bacillus circulans *(Bc) (spore), and* Bacillus subtilis *(Bs) (spore) treated in McIlvaine buffer pH 7.0 with MS (0.2 MPa, 40°C, 450 W and 4 minutes, for spores 15 minutes*). Adapted from [31, 45].*

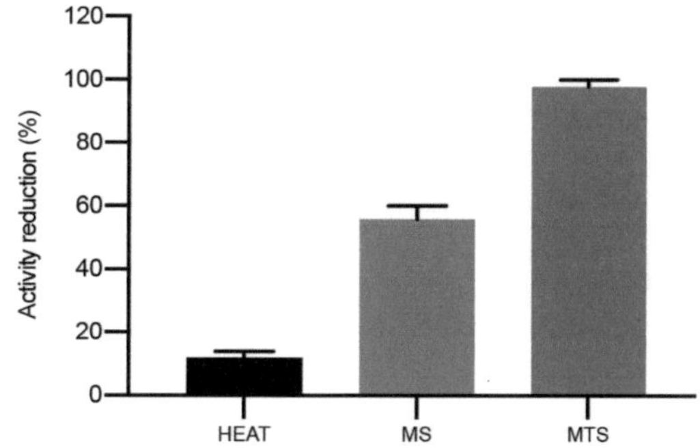

Figure 3.
Activity reduction of pectin methylesterase of tomato juice treated by heat, MS, and MTS treatments at 62.5°C and 1 minute (ultrasonic conditions: 20 kHz, 750 W, 0.2 MPa). Adapted from [40].

the treatment medium, the same levels of decontamination may not be achieved throughout the whole material or surface [47].

In the food industry, ultrasonic baths can also be used to clean and decontaminate surfaces of products such as vegetables, fruit, eggs, fish, etc., but always bear in mind that in the best scenario, a microbial inactivation of 1 Log_{10} cycle (90% reduction of the microbial population) could be achieved. Based on this, in meat industry, water-steam-based-systems combined with ultrasound have been recently proposed for poultry carcasses decontamination [48]. Thus, Boysen and Rosenquist [49] studied the inactivation of *Campylobacter* from broiler skins after applying different physical decontamination methods. They observed that steam-ultrasound was the most effective method achieving an inactivation of 2.5 Log_{10} reductions, 1 Log_{10} extra-reduction compared with other systems. However, the carcasses appeared to be slightly boiled after the treatment. Musavian et al. [50] decontaminated broiler carcasses with ultrasound (30–40 kHz) and steam (90–94°C) combination and observed additional reduction of 1–1.4 Log_{10} cycles of *Campylobacter* after applying 10 s of treatment. An example of this application is the SonoSteam system [51].

Regarding the cleaning of equipment surfaces, a widely known example in the food industry is the application of ultrasound for cleaning wine-aging barrels. It allows an effective cleaning even within the wood pores where spoilage microorganisms such as *Brettanomyces* are located, since ultrasound can remove part of the layers created by the precipitation of crystallized tartrates [52]. The additional advantage of this effect is that the aroma of the oak is maintained, reducing maintenance costs and the need to replace the barrels [53].

Finally, one of the most recent applications in terms of cleaning has been the use of ultrasound for the disintegration of bacterial biofilms generated on working surfaces of the food industry that can lead to cross-contaminant phenomena. Thus, the use of ultrasound would allow to reduce the formation or even to eliminate these biofilms, for example, in conveyor belts used for the transportation of foods inside the industry [54]. An industrial example of this application has been developed by Lubing systems [55].

Besides the ultrasound-assisted microbial decontamination, a recent study has demonstrated the potential of ultrasound for reducing the heavy metal load from foods. Condón-Abanto et al. [56] observed that the cadmium content of edible crabs (*Cancer pagurus*) was reduced by 23% after their immersion in water at 50°C for 40 minutes applying ultrasound (35 kHz, 200 W). The same treatment without ultrasound scarcely reduced the Cd content of 2%. These results open the possibility of reducing chemical contaminants or other chemical risks present in foods by using ultrasound as it will be discussed later on.

5.3 Mass transfer

The processes of mass transfer between two phases consist of the transfer of a certain component from one phase to another as a result of the difference in concentration between both phases. In the food industry, mass transfer occurs in many processes, such as the extraction of compounds of interest from inside the cells of a food product (sucrose, colorants, etc.), the elimination of water in processes like drying/dehydration, or the incorporation of solids as it happens when marinating and/or pickling.

5.3.1 Extraction

The traditional method for the extraction of intracellular compounds of interest for the food industry (sugar, colorants, bioactive substances such as polyphenols, etc.) consists on using an adequate solvent combined with other systems such as heat, agitation, etc. However, this technique has some disadvantages such as the high electrical consumption—becoming up to 70% of the required energy to extract a certain compound—high water requirements, and the use of toxic or contaminant solvents. For this reason, the food industry has struggled to find more profitable and eco-friendly methods for the extraction of compounds [16, 57], such as ultrasound, which improves the extraction efficiency by applying lower temperatures and shorter processing times than traditional extracting methods [58].

The extraction of aromatic compounds, antioxidants, pigments, and other organic or inorganic substances from tissues, mostly vegetal, has been widely investigated and successfully carried out by applying high-power ultrasound [59–63]. The application of ultrasound to a vegetable product immersed in a liquid medium can induce rapid fragmentation of the material, increasing the surface area of the solid in contact with the solvent and accelerating the mass transfer and, therefore, the extraction rate and yield [64]. Several advantages have been pointed out for the ultrasound-assisted extraction including the reduction of extraction time, energy, and the amount of solvent used and of unit operations and also a rapid return of investment [57]. As a way of example, **Figure 4** shows the extraction yield of chlorophyll from spinach leaves by using or not ultrasound (20 kHz) [57]. As observed, the amount of chlorophyll extracted was fourfold higher than in the control process after 20 minutes of maceration using ultrasound and more than double than the control after 80 minutes of extraction.

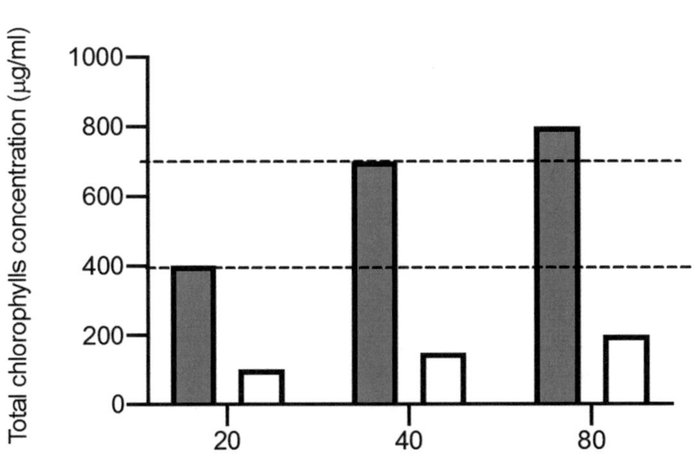

Figure 4.
Total chlorophylls concentration (µg/ml) extracted from spinach leaves treated (filled bars) or not (white bars) with ultrasound (20 kHz). Adapted from [57].

Besides the recovery of compounds of interest, also the extraction of potential risky compounds for human health is under investigation like oligosaccharides from pulses or Cd from edible crabs [65, 56]. In the same direction, the use of ultrasound has been recently evaluated for reducing the acrylamide content of fried potatoes which is a carcinogenic compound. By applying a pre-frying treatment of 30 minutes by immersing potatoes in an ultrasound water bath at 35 kHz, 92.5 W/kg, and 42°C, Antunes-Rohling et al. [66] obtained a 90% reduction in acrylamide compared to potatoes directly fried and a 50% reduction compared to potatoes soaked in water but with no ultrasound applied.

Based on the showed possibilities of ultrasound for extracting compounds of interest, different semi-industrial systems have been developed which are detailed in the revision of Chemat et al. [16]. More recently, and based on the works done in the winery industry, a continuous ultrasound system has been constructed in order to improve the extraction of polyphenolic compounds from grapes [67]. Wine is a product highly appreciated for its organoleptic properties such as color, aroma, and flavor. The application of ultrasound has been studied in the wine maceration process to favor the extraction of polyphenols responsible for color [68] and in the lees (*aging on lees*) for the extraction of polysaccharides responsible for color stability, mouthfeel, and reduction of wine's astringency [69].

5.3.2 Drying and dehydration

In the food industry, drying and dehydration of foods are important preserving processes where mass and energy transfer phenomena occur. They consist of removing a large part of the water from the food in order to improve the stability of the product, reducing its volume and weight and facilitating the handling and transport of the products [70, 71]. Currently, one of the most widespread techniques in the food industry is air convection dehydration. However, it is an energetically costly operation and, in some cases, requires long periods of time. In order to reduce drying times, some industrial strategies exist, such as increasing the temperature of the air, which can cause alterations in the composition and structure of foods, or increasing the air speed that might lead to the formation of a dry and impermeable layer that can inhibit the exit of humidity from the interior of the product [70].

Ultrasound has been evaluated as an alternative to traditional dehydration systems. In this case, the water removal process is improved mainly by the phenomenon known as "sponge effect" which enhances the diffusion of water from the interior of the product to the surface [72]. Nonetheless, cavitation of intracellular and extracellular water may also occur, forming new microchannels [73]. In addition, the application of ultrasound through the air generates turbulence that produces an important microstreaming at the interface between food and air which help remove surface moisture [74].

Ultrasound-assisted dehydration in food has been researched since the 1950s and 1960s, but it has been in recent years when major advances have been made since new family of piezoelectric transducers with extensive radiating surface have been developed [75]. There are two types of ultrasound application systems in food dehydration processes: by direct contact between the transducer and the food and by indirect contact through the air (airborne ultrasound systems). Contact systems, even though they are more efficient, can cause product damage, equipment development is complicated, and specific hygiene requirements are necessary. In any case, very promising results were obtained by De la Fuente-Blanco et al. [72], drying carrot cylinders achieving a faster loss of water than the usual dehydration by forced air process and obtaining a final moisture content in the product of less than 1%.

More studies have been carried out with airborne ultrasound systems, reducing drying times by 20–30% when applied at low temperatures and low air velocities [70]. For example, García-Pérez et al. [76] developed a convection drying equipment applying ultrasound to the air, in which the treatment chamber consisted of a vibratory aluminum cylinder coupled to a transducer (21.8 kHz, 75 W). In this study, they achieved a reduction of 26.7% in drying time of carrot skin samples when dried at 40°C and 0.6 m/s. The effect of air temperature (30–70°C) on the speed of drying with ultrasound was demonstrated by the same authors [77]. They obtained an increase in diffusion coefficient of 23.6% at 30°C, while at 70°C only 1.3%. These studies indicate that at either high air velocities or high temperatures, the effects of these parameters predominate over ultrasound.

In addition to improving convection drying processes, studies have also been carried out on the application of ultrasound in vacuum drying [78, 79] or in freeze-drying [80, 81] obtaining higher drying rates than the traditional process.

As it can be appreciated, the obtained results are promising; however, at present, pilot or industrial systems are scarce. The main technological challenges to address are basically reducing the overheating produced by the transducers and adapting the frequency and ultrasonic power to the working conditions, taking into account the acoustic impedance, attenuation, and absorption of the product to be dehydrated [73].

5.3.3 Marinating and pickling

Marinating and pickling are food preservation techniques used in vegetables, meat, and fish products. Brine, vinegar, or other organic acids; oil; and spices are usually used. In general, long processes are required, which involves the immobilization of the product resulting in economic costs and also potentially leading to structural damage, softening, and swelling, which might affect the quality of the product [16]. The application of high-power ultrasound between frequencies of 20 and 50 kHz has made possible to shorten pickling or brine contact times. Besides, in the case of meat such as pork loin, the water and salt content of the samples was increased (63–65% and 7–50%, respectively) when ultrasound (20 kHz, >39 W/cm^2) was applied compared to brining in static mode and with mechanical agitation. With an intensity higher than 64 W/cm^2, the water content of the samples after the process was even higher than that of fresh meat [82]. Improvement of water intake has been observed also in fish. Thus, a 6% higher water intake of thawed cod fillets after 48 hours of hydration than the standard process when applying ultrasound (40 kHz, 3.9 W/kg) was observed [10].

5.4 Energy transfer

Energy transfer (e.g., heating or freezing) together with mass transfer are common unit operations in the food industry. Both direct and indirect applications of ultrasound have been used to increase the energy transfer rates of traditional heating/freezing systems. Ultrasonic waves produce a direct heating of the product/medium due to the great energy released in the medium, as well as an intense agitation favoring a faster and more uniform heating of the product. On the other hand, the vibration caused by the indirect application of ultrasound accelerates the transfer of heat from traditional systems, both to release it in cooling processes and to provide it when heating.

5.4.1 Heating

The use of ultrasound to improve heating of liquid and solid foods is known since the 1960s [83]. However, scarce scientific information has been published till recent years. It has been described that ultrasound (20 kHz, 75 W) can increase the conductive heat transfer when applied in metals by 2.3- and 5.5-folds [83], becoming this effect the basis of the design of heat exchangers including ultrasound systems [83, 84]. In the case of liquid foods, the application of ultrasound of 20 kHz also improved the convection heat transfer in this case up to 25-fold in water [85]. In the case of viscous liquids such as puree, creams or soups, ultrasound not only improved the energy transfer but also the uniformity of the heating. Thus, an increase in energy transfer of 33 and 43% when heating tomato soup assisted with 45 and 450 W of ultrasound (20 kHz), respectively, was observed (**Figure 5**).

Finally, the application of ultrasound in hot water to heat solid products resulted in a faster heat transfer, reducing the time to apply pasteurization treatments or even to cook food products and therefore getting higher quality products [16].

Some authors have studied the improvement of heating for food cooking by using ultrasound. One of the first studies was conducted by Pohlman et al. [86] who evaluated the effects of ultrasound for cooking different pieces of beef. An ultrasonic field of 22 W/cm^2 was applied and compared to the traditional cooking of beef in a convection oven up to 70°C in the center. Ultrasonic cooking reduced the cooking time by 54% and the energy consumption of the process by 42%. In addition, samples cooked with ultrasound were cooked more uniformly and showed higher water retention, lower cooking losses, and lower hardness. In recent years, more studies have been conducted on this topic. More specifically, ultrasound has been used to accelerate heat transfer in the pasteurization of packaged sausages [87] and of ready-to-eat whole brown crab [88], to evaluate the frying-assisted ultrasound process of meatballs [89] and for the cooking of mortadella [90]. Even more, improvements in heat transfer have been observed at boiling water temperatures and over atmospherically pressure. Thus, 20% and up to 32% reduction in the cooking times were observed when boiling macaroni at 100°C or chickpeas at 120°C and 0.09 MPa, respectively, in an ultrasonic field of 40 kHz and 25 W/kg by using a new patented ultrasound system [91, 92].

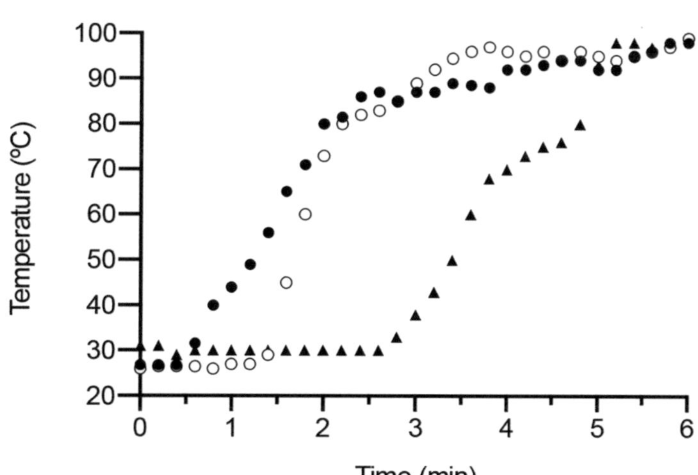

Figure 5.
Evolution of the temperature during the heating of tomato soup at different ultrasound (20 kHz) intensities: 0 (▲), 45 (○), and 450 W (●).

In summary, the application of ultrasound allows to reduce the heating times by enhancing the energy transfer in liquid, viscous, and solid products and applying more uniform thermal treatments reducing the number of cold spots.

5.4.2 Freezing

Freezing is one of the oldest methods for food preservation. It involves subjecting food to temperatures lower than that of the freezing point causing the conversion of food water into ice and thereby limiting microbial growth and chemical and enzymatic reactions. When freezing speed is slow, large crystals with edges are formed in the extracellular liquid, causing the loss of water from inside the cells. This leads to dehydration, cell contraction, and partial plasmolysis; these phenomena, together with the damage caused by ice crystals that cause injuries in cell membranes, lead to water leakage after defrosting, producing the loss of food quality. On the other hand, quick freezing produces small ice crystals in the intracellular and extracellular space, resulting in less cell damage and in higher quality products [93]. Ultrasound-assisted freezing reduces treatment time by favoring both nucleation and controlled crystal growth [16]. These effects have been mainly attributed to acoustic cavitation and the microstreaming generated in the liquid as well as the microbubbles that act as nuclei of crystallization [94]. **Figure 6** shows the freezing curves of 2 cm × 2 cm cylinders of meat sausages frozen in an ultrasound bath at −22°C applying or not ultrasound (40 kHz, 50 W). As can be observed, application of ultrasound reduced the freezing time and even eliminated the water-ice crystal transition phase.

Several studies have been carried out on the application of ultrasound during the freezing process of foods. In most of these studies, ultrasound has been applied using ultrasound baths with the product immersed in an aqueous medium, e.g., panaria dough [95], potatoes [96], broccoli [97], apples [98], mushrooms [99], and pork loin [100]. For example, Sun et al. [101] studied the influence of ultrasound-assisted immersion freezing on the process and on the quality of common carp (*Cyprinus carpio*). The application of ultrasound at 30 kHz and 175 W reduced the freezing time of 37.2%, being this ultrasound intensity the optimal, since below it the effect of ultrasound was undetectable and above it overheating was observed due to the high ultrasound intensities applied. This increase in the freezing rate

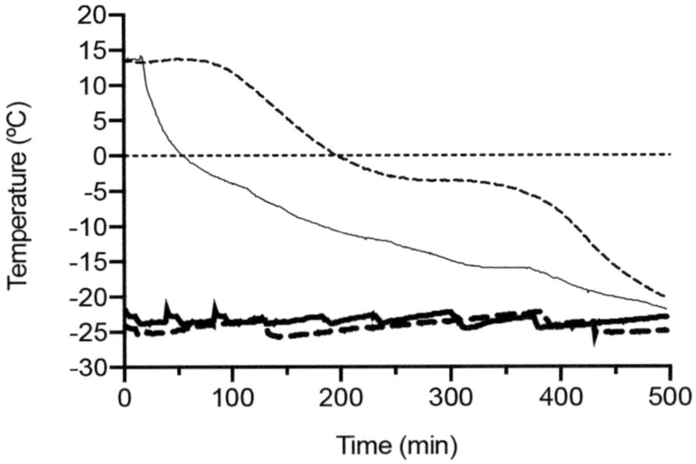

Figure 6.
Temperature of sausages (thin lines) and media (thick lines) when applying (continuous lines) or not ultrasound (40 kHz, 50 W/kg) when freezing.

resulted in an improvement in the product quality since the cooking loss (% of loss water after cooking the product) values determined were similar to those of fresh product: 7.9% in fresh product *vs* 8.3% when ultrasound was applied.

5.4.3 Thawing

Thawing is as important as freezing in the food industry, since a large proportion of frozen foods require thawing prior to their consumption. This process must be carried out as quick as possible to avoid affecting the hygienic quality of the product, but bearing in mind that the higher the speed, the worse will be the sensory characteristics of the final product because time is required for the cells to reabsorb the released water during freezing. As it has been explained, the application of ultrasound would help to improve the transfer of energy due to the cavitation and the microstreaming generated in the liquid [96]. Some studies have been carried out in beef, pork, and codfish [102]; pork *Longissimus dorsi* muscle [103]; and tuna [104]. In a study carried out by Gambuteanu and Alexe [103], thawing assisted by ultrasound in samples of pork *Longissimus dorsi* muscle was evaluated. Experiments were performed at intensities of 0.2, 0.4, and 0.6 W/cm^2 in a water bath at 15°C, and they were compared with thawing in air at 15°C and thawing by immersion on water at 15°C. The thawing rate was influenced by the intensity of ultrasound treatment: the higher the ultrasonic intensity, the shorter the thawing time. Thus, the thawing rates for air and water immersion were 0.16 and 0.29°C/min, respectively, whereas, for ultrasound intensities of 0.2, 0.4, and 0.6 W/cm^2, the values were 0.62, 0.73, and 1°C/min. Therefore, the thawing time of pork samples could be reduced applying ultrasound technology. Similar conclusions were obtained by our research group in cod fillets thawed in an ultrasound water bath at 2°C (25 kHz, 14.7 W/kg), reducing 65% the time to achieve 0°C, maintaining the water holding capacity and cook loss of the fresh product, and with a better sensorial quality than the air defrosted product (**Figure 7**).

5.5 Other applications of ultrasound in the food industry

In addition to the applications already described, ultrasound technology has been evaluated and applied in the food industry to improve other processes whose result is based mainly on mechanical effects.

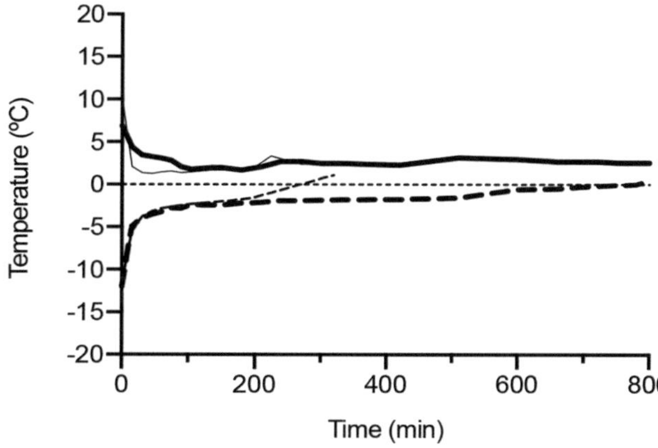

Figure 7.
Temperature of cod fillets (thin lines) and water (think lines) when applying (continuous lines) or not ultrasound (25 kHz, 14.7 W/kg) when thawing.

5.5.1 Foaming and degassing capacity

Foam is a dispersion of gas in a liquid medium that is often formed during the manufacture of many products, as a result of aeration or agitation of liquids, during vaporization of liquids, or due to chemical or biological reactions [105]. Mechanical methods are the most effective at removing unwanted foams during food processing, compared to antifoaming chemical agents. The use of ultrasound can be considered a mechanical method of foam removal, since it is based on the propagation of the sound waves through the foam, without affecting the liquid [106]. For this application, airborne transducers are mainly used [107].

Another increasingly widespread application of ultrasound is degassing. Liquids contain dissolved gases such as oxygen, carbon dioxide, or nitrogen. Conventionally, to degas a liquid, it is boiled or subjected to vacuum, reducing the solubility of the gases. Ultrasonic degassing has the advantage of not substantially increasing the temperature of the liquid. In the presence of an ultrasonic field, the gas bubbles begin to vibrate, coalesce, and grow, reaching a sufficient size to ascend to the liquid surface, being thus removed from the aqueous medium [16].

In a study that covers both applications, foam removal and degassing, Villamiel et al. [108] used 1-second ultrasonic pulses (20 kHz) in milk. At 20°C with 3 minutes of treatment, they managed to reduce foam by 80% with an energy consumption of 40 kJ/l. In order to eliminate the oxygen dissolved in milk, a more energetic treatment was necessary (240 kJ/l).

5.5.2 Filtration

High-power ultrasound has been applied to promote diffusion through membranes and porous materials. This improvement is attributed to the formation of microstreams generated within the liquid in the presence of high-energy ultrasonic fields. That is how it would facilitate the processes of filtration, ultrafiltration, dialysis, and reverse osmosis [109]. During membrane filtration, the flow progressively decays to a stationary state due to the polarization of the concentration, and the saturation of the filter. Ultrasound acts by increasing the flow and preventing saturation if applied during filtration or by breaking the deposit layer of solutes or cake on the filters, acting in this case as a cleaning method [16].

5.5.3 Texture modification

Texture plays a crucial role in influencing consumers' liking and preference of meat products. This sensation is influenced by various factors including muscle type, age and cut, its water holding capacity, and the degree of maturation, among others [110]. The application of ultrasound might help to improve meat tenderness, thus obtaining better quality products. However, the effect of high-power ultrasound on meat tenderization is not entirely clear, and this is likely because there are many factors that influence its effect, such as the characteristics of the ultrasonic field, the time of exposure, the animal species, and the type of muscle, among others. Some authors state that those studies in which ultrasound application had no effect would be due to the low ultrasonic densities (0.29–1.55 W/cm^2) or short treatment times (15 s) applied [111–113]. In any case, there are systems already in the market for tendering meat based on ultrasound [114].

In the case of meat products, ultrasound can improve cohesiveness between different pieces of meat [109] by promoting the release of myofibrillar proteins and gel formation. This effect is important in processes such as the production of cooked ham or cured meat products in which an adhesive protein exudate is required in order to act as a glue between the different parts during molding or stuffing [110].

5.5.4 Food cutting

Most processed foods are prepared in large quantities, often in blocks or in large sheets. For marketing and consumption, it is necessary to reduce their size, in many cases by cutting the product. For this propose, ultrasonic probes in the shape of a blade are used which vibrate at a certain ultrasonic frequency longitudinally or as a piston. When it comes into contact with food, it cuts it due to both the vibration and the sharp edge of the blade. These types of probes have been used successfully in the cutting of fragile, heterogeneous, and sticky products such as cream cakes, bread, pastries, biscuits, and cheese [16, 115].

6. Conclusions

Although ultrasound is a well-known technology that is commonly used in several fields such as medicine or in the automobile industry, its use in the food industry is still scarce especially in the case of high-power ultrasound. However, due to its capacity to improve mass and energy transfer phenomena—which occur in numerous processes in the food industry—it might be very helpful for producing safer and higher quality products than those obtained by traditional procedures. In addition, ultrasound is considered a safe, nonpolluting, and environmentally friendly technology, which has also contributed to attract the interest of the food industry. Finally, the lower implementation cost—up to the industrial scale—of some applications compared to other nonthermal technologies such as pulsed electric fields or high hydrostatic pressures will facilitate its industrialization in some food sectors. In any case, further research is still necessary for some applications since many factors have to be considered when designing equipment and applying ultrasound treatments in the food industry in order to achieve an efficient application.

Acknowledgements

The authors wish to acknowledge the financial support from iNOBox (Project number 281106) funded by the Research Council of Norway and the Department of Innovation Research and University of the Aragon Government and European Social Fund (ESF). L.A. gratefully acknowledges the financial support for her studies provided by the "Ministerio de Educación y Formación Profesional."

Author details

Leire Astráin-Redín, Salomé Ciudad-Hidalgo, Javier Raso, Santiago Condón, Guillermo Cebrián and Ignacio Álvarez*
Departamento de Producción Animal y Ciencia de los Alimentos, Tecnología de los Alimentos, Facultad de Veterinaria, Instituto Agroalimentario de Aragón (IA2), Universidad de Zaragoza-CITA, Zaragoza, Spain

*Address all correspondence to: ialvalan@unizar.es

References

[1] Gallego-Juárez JA. Basic principles of ultrasound. In: Villamiel M et al., editors. Ultrasound in Food Processing. Recent Advances. Chichester: Wiley Blackwell; 2017. pp. 4-26

[2] Hecht E. Physics: Calculus. Pacific Grove, CA: Brooks/Cole; 1996. pp. 445-521

[3] Kentish S, Ashokkumar M. The physical and chemical effects of ultrasound. In: Feng H, Barbosa-Cánovas GV, Weiss J, editors. Ultrasound Technologies for Food and Bioprocessing. London: Springer; 2011. pp. 1-13

[4] Zupanc M, Pandur Z, Perdih TS, Stopar D, Petkovsek M, Dular M. Effects of cavitation on different microorganisms: The current understanding of the mechanisms taking place behind the phenomenon. A review and proposals for further research. Ultrasonics Sonochemistry. 2019;**57**:147-165. DOI: 10.1016/j.ultsonch.2019.05.009

[5] Suslick K. Sonochemistry. Science. 1990;**247**:1439-1445

[6] Floros JD, Liang H. Acoustically assisted diffusion through membranes and biomaterials. Food Technology. 1994;**48**:79-84

[7] Mulet A, CárceL JA, Benedito J, Roselló C, Simal S. Ultrasonic mass transfer enhancement in food processing. In: Welti-Chanes J, Vélez-Ruiz J, Barbosa-Canova G, editors. Transport Phenomena in Food Processing. New York: CRC Press; 2003

[8] Mulet A, Cárcel JA, Sanjuán N, García-Pérez JV. Food dehydration under forced convection conditions. In: Delgado J, editor. Recent Progress in Chemical Engineering. Houston: Studium Press LLC; 2010

[9] Mason TJ et al. Other non-thermal processing techniques: Application of ultrasound. In: Sun DW, editor. Emerging Technologies for Food Processing. London: Academic Press; 2005. pp. 323-345

[10] Antunes-Rohling A, Raso J, Cebrián G, Álvarez I. Ultrasound technology to reduce technological adjuvant when hydrating thawed cod fillets. In: Proceedings the IFT-EFFoST 2018 International Nonthermal Processing Workshop and Short Course. Sorrento; 2018

[11] Kentish S. Engineering principles of ultrasound technology. In: Bermúdez-Aguirre D, editor. Ultrasound: Advances in Food Processing and Preservation. London: Academic Press; 2017

[12] Povey MJW, Mason TJ. Ultrasound in Food Processing. London: Blackie Academic and Professional; 1998

[13] Bermúdez-Aguirre D, Mobbs T, Barbosa-Cánovas GV. Ultrasound applications in food processing. In: Feng H, Barbosa-Cánovas GV, Weiss J, editors. Ultrasound Technologies for Food and Bioprocessing. New York: Springer; 2011. pp. 65-106

[14] Saclier M, Peczalski R, Andrieu J. Effect of ultrasonically induced nucleation on ice crystals' size and shape during freezing in vials. Chemical Engineering Science. 2010;**65**:3064-3071. DOI: 10.1016/j.ces.2010.01.035

[15] Beck SM, Sabarez H, Gaukel V, Knoerzer K. Enhancement of convective drying by application of airborne ultrasound—A response surface approach. Ultrasonics Sonochemistry. 2014;**6**:2144. DOI: 10.1016/j.ultsonch.2014.02.013

[16] Chemat F, Huma Z, Khan MK. Applications of ultrasound in food

technology: Processing, preservation and extraction. Ultrasonics Sonochemistry. 2011;**18**:813-835. DOI: 10.1016/j.ultsonch.2010.11.023

[17] Charoux C, Ojha S, O'Donnell C, Cardoni A, Brijesh T. Applications of airborne ultrasonic technology in the food industry. Journal of Food Engineering. 2017;**208**:28-36. DOI: 10.1016/j.jfoodeng.2017.03.030

[18] Li MK, Fogler HS. Acoustic emulsification. Part I. The instability of the oil-water interface to form the initial droplets. Journal of Fluid Mechanics. 1978;**88**:499-511

[19] Li MK, Fogler HS. Acoustic emulsification. Part II. Breakup of the primary oil droplets in a water medium. Journal of Fluid Mechanics. 1978;**88**:513-528

[20] Cucheval A, Chow RCY. A study on the emulsification of oil by power ultrasound. Ultrasonics Sonochemistry. 2008;**15**:916-920. DOI: 10.1016/j.ultsonch.2008.02.004

[21] Thompson LH, Doraiswamy LK.Sonochemistry: Science and engineering. Industrial Engineering and Chemical Research. 1999;**38**:1215-1249

[22] Abismaïl B, Canselier JP, Wilhelm AM, Delmas H, Gourdon C. Emulsification by ultrasound: Drop size distribution and stability. Ultrasonics Sonochemistry. 1999;**6**:75-83

[23] Carpenter J, Saharan VK. Ultrasonic assisted formation and stability of mustard oil in water nanoemulsion: Effects of process parameters and their optimization. Ultrasonics Sonochemistry. 2017;**35**:422-430. DOI: 10.1016/j.ultsonch.2016.10.021

[24] Jafari SM, He Y, Bhandari B. Production of sub-micron emulsions by ultrasound and microfluidification techniques. Journal of Food Engineering. 2007;**82**:478-488

[25] Ramachandran KB, Sulaiman AZ, Fong C, Gak C, et al. Kinetic study of hydrolysis of oils by lipase with ultrasonic emulsification. Biochemical Engineering Journal. 2006;**32**:19-24. DOI: 10.1016/j.bej.2006.08.012

[26] Ashokkumar M. Applications of ultrasound in food and bioprocessing. Ultrasonics Sonochemistry. 2015;**25**:17-23. DOI: 10.1016/j.ultsonch.2014.08.012

[27] Ghasemi S, Abbasi S. Formation of natural casein micelle nanocapsule by means of pH changes and ultrasound. Food Hydrocolloids. 2014;**42**:42-47. DOI: 10.1016/j.foodhyd.2013.10.028

[28] Awad TS, Moharram HA, Shaltout OE, Asker D, Youssef MM. Applications of ultrasound in analysis, processing and quality control of food: A review. Foodservice Research International. 2012;**48**:410-427. DOI: 10.1016/j.foodres.2012.05.004

[29] Gaboriaud PLF. Sterilisation des liquides par ultrasons. French Patent. 1984;**2**:575-641

[30] Jacobs SE, Thonrley MJ. The lethal action of ultrasonic waves on bacteria suspended in milk and other liquids. The Journal of Applied Bacteriology. 1954;**17**:38-56

[31] Pagán R, Mañas P, Raso J, Condón S. Bacterial resistance to ultrasonic waves under pressure at nonlethal (manosonication) and lethal (manothermosonication) temperatures. Applied and Environmental Microbiology. 1999;**65**:297-300

[32] Arroyo C, Lyng JG. The use of ultrasound for the inactivation of microorganisms and enzymes. In: Villamiel M et al., editors. Ultrasound in Food Processing. Recent Advances.

Chichester: Wiley Blackwell; 2017. pp. 258-278

[33] Condón S, Mañas P, Cebrián G. Manothermosonication for microbial inactivation. In: Feng H, Barbosa-Cánovas GV, Weiss J, editors. Ultrasound Technologies for Food and Bioprocessing. New York: Springer; 2011. pp. 287-320

[34] Alzamora SM, Guerrero SN, Schenk M, Raffellini S, López-Malo A. Inactivation of microorganisms. In: Feng H, Barbosa-Cánovas GV, Weiss J, editors. Ultrasound Technologies for Food and Bioprocessing. Nueva York: Springer; 2011. pp. 321-344

[35] Piyasena P, Mohareb E, McKellar RC. Inactivation of microbes using ultrasound: A review. International Journal of Food Microbiology. 2003;**87**:207-216. DOI: 10.1016/s0168-1605(03)00075-8

[36] Huang Q, Li L, Fu X. Ultrasound effects on the structure and chemical reactivity of cornstarch granules. Starch-Stärke. 2007;**59**:371-378. DOI: 10.1002/star.200700614

[37] Walkling-Ribeiro M, NocI F, Cronin DA, Lyng JG. Shelf life and sensory evaluation of orange juice after exposure to thermosonication and pulsed electric fields. Food and Bioproducts Processing. 2009;**87**:102-107. DOI: 10.1016/j.fbp.2008.08.001

[38] Raso J, Pagan R, Condon S, Sala FJ. Influence of temperature and pressure on the lethality of ultrasound. Applied and Environmental Microbiology. 1998;**64**:465-471

[39] López P et al. Inactivation of peroxidase, lipoxygenase, and polyphenol oxidase by manothermosonication. Journal of Agricultural and Food Chemistry. 1994;**42**:252-256

[40] Lopez P, Vercet A, Sanchez AC, Burgos J. Inactivation of tomato pectic enzymes by manothermosonication. Zeitschrift für Lebensmitteluntersuchung und-Forschung A. 1998;**207**:249-252

[41] Mawson R, Gamage M, Terefe NS, Knoerzer K. Ultrasound in enzyme activation and inactivation. In: Feng H, Barbosa-Cánovas GV, Weiss J, editors. Ultrasound Technologies for Food and Bioprocessing. London: Springer; 2011. pp. 369-404

[42] Terefe NS, Buckow R, Versteeg C. Quality-related enzymes in plant-based products: Effects of novel food-processing technologies part 3: Ultrasonic processing. Critical Reviews in Food Science and Nutrition. 2015;**55**:147-158. DOI: 10.1080/10408398.2011.586134

[43] Vercet A, Burgos J, Lopez-Buesa P. Manothermosonication of heat- resistant lipase and protease from *Pseudomonas fluorescens*: Effect of pH and sonication parameters. The Journal of Dairy Research. 2002;**69**:243-254. DOI: 10.1017/s0022029902005460

[44] Vercet A, Lopez P, Burgos J. Inactivation of heat-resistant pectinmethylesterase from orange by manothermosonication. Journal of Agricultural and Food Chemistry. 1999;**47**:432-437

[45] Sala FJ, Burgos J, Condon S, Lopez P, Raso J. Effect of heat and ultrasound on microorganisms and enzymes. In: Gould GW, editor. New Methods of Food Preservation. London: Blackie Academic and Professional; 1995. pp. 176-204

[46] Mason TJ. Ultrasonic cleaning: An historical perspective. Ultrasonics Sonochemistry. 2016;**29**:519-523. DOI: 10.1016/j.ultsonch.2015.05.004

[47] Zhou B, Lee H, Feng H. Microbial decontamination of food by power ultrasound. In: Demirci A, Ngadi OM, editors. Woodhead Publishing Series in Food Science, Technology and Nutrition. Philadelphia, PA, USA: Woodhead Publishing; 2012. pp. 300-321

[48] Turantaş F, Kılıç G, Kilic B. Ultrasound in the meat industry: General applications and decontamination efficiency. International Journal of Food Microbiology. 2015;**198**:59-69. DOI: 10.1016/j.ijfoodmicro.2014.12.026. 2015

[49] Boysen L, Rosenquist H. Reduction of thermotolerant campylobacter species on broiler carcasses following physical decontamination at slaughter. Journal of Food Protection. 2009;**72**:497-502. DOI: 10.4315/0362-028X-72.3.497

[50] Musavian HS, Krebs NH, Nonboe U, Corry JEL, Purnell G. Combined steam and ultrasound treatment of broilers at slaughter: A promising intervention to significantly reduce numbers of naturally occurring campylobacters on carcasses. International Journal of Food Microbiology. 2014;**176**:23-28. DOI: 10.1016/j.ijfoodmicro.2014.02.001

[51] Available from: https://sonosteam.com [Accessed: October 2019]

[52] Porter G, Lewis A, Barnes M, Williams R. Evaluation of high power ultrasound porous cleaning efficacy in American oak wine barrels using X-ray tomography. Innovative Food Science & Emerging Technologies. 2011;**12**:509-514. DOI: 10.1016/j.ifset.2011.06.007

[53] Available from: https://cavitus.com/?lang=es [Accessed: October 2019]

[54] Fink R, Oder M, Stražar E, Filip S. Efficacy of cleaning methods for the removal of *Bacillus cereus* biofilm from polyurethane conveyor belts in bakeries. Food Control. 2017;**80**:267-272. DOI: 10.1016/j.foodcont.2017.05.009

[55] Available from: https://www.lubing.com/new-compact-cleaning-unit.html [Accessed: October 2019]

[56] Condón-Abanto S et al. Evaluation of the potential of ultrasound technology combined with mild temperatures to reduce cadmium content of edible crab (*Cancer pagurus*). Ultrasonics Sonochemistry. 2018;**48**:550-554. DOI: 10.1016/j.ultsonch.2018.07.019

[57] Chemat F et al. Ultrasound assisted extraction of food and natural products. Mechanisms, techniques, combinations, protocols and applications. A review. Ultrasonics Sonochemistry. 2017;**34**:540-580. DOI: 10.1016/j.ultsonch.2016.06.035

[58] Patist A, Bates D. Ultrasonics innovations in the food industry: From the laboratory to commercial production. Innovative Food Science & Emerging Technologies. 2008;**9**:147-154. DOI: 10.1016/j.ifset.2007.07.004

[59] Luengo E, Condón-Abanto S, Condón S, Álvarez I, Raso J. Improving the extraction of carotenoids from tomato waste by application of ultrasound under pressure. Separation and Purification Technology. 2014;**136**:130-136. DOI: 10.1016/j.seppur.2014.09.008

[60] Karki B et al. Enhancing protein and sugar release from defatted soy flakes using ultrasound technology. Journal of Food Engineering. 2010;**96**:270-278. DOI: 10.1016/j.jfoodeng.2009.07.023

[61] Xie P et al. Enhanced extraction of hydroxytyrosol, maslinic acid and oleanolic acid from olive pomace: Process parameters, kinetics and thermodynamics, and greenness assessment. Food Chemistry.

2019;**276**:662-674. DOI: 10.1016/j.foodchem.2018.10.079

[62] Menezes Maciel Bindes M, Hespanhol Miranda Reis M, Luiz Cardoso V, Boffito DC. Ultrasound-assisted extraction of bioactive compounds from green tea leaves and clarification with natural coagulants (chitosan and Moringa oleífera seeds). Ultrasonics Sonochemistry. 2019;**51**:111-119. DOI: 10.1016/j.ultsonch.2018.10.014

[63] Chen C et al. Ultrasound-assisted extraction from defatted oat (*Avena sativa* L.) bran to simultaneously enhance phenolic compounds and β-glucan contents: Compositional and kinetic studies. Journal of Food Engineering. 2018;**222**:1-10. DOI: 10.1016/j.jfoodeng.2017.11.002

[64] Mason TJ, Vinatoru M. Ultrasonically assisted extraction in food processing and the challenges of integrating ultrasound into the food industry. In: Villamiel M et al., editors. Ultrasound in Food Processing. Recent Advances. Chichester: Wiley; 2017. pp. 329-353

[65] Han IH, Baik B. Oligosaccharide content and composition of legume and their reduction by soaking, cooking, ultrasound and high hydrostatic pressure. Cereal Chemistry. 2006;**83**:428-433. DOI: 10.1094/CC-83-0428

[66] Antunes-Rohling A et al. Ultrasound as a pretreatment to reduce acrylamide formation in fried potatoes. Innovative Food Science & Emerging Technologies. 2018;**49**:58-169. DOI: 10.1016/j.ifset.2018.08.010

[67] Maza MA, Álvarez I, Raso J. Thermal and non-thermal physical methods for improving polyphenol extraction in red winemaking. Beverages. 2019;**5**:47. DOI: 10.3390/beverages5030047

[68] El Darra N, Grimi N, Maroun R, Louka N, Vorobiev E. Pulsed electric field, ultrasound, and thermal pretreatments for better phenolic extraction during red fermentation. European Food Research and Technology. 2012;**236**:47-56. DOI: 10.1007/s00217-012-1858-9

[69] Del Fresno JM et al. Application of ultrasound to improve lees ageing processes in red wines. Food Chemistry. 2018;**261**:157-163. DOI: 10.1016/j.foodchem.2018.04.041

[70] Cárcel J, Castillo D, Simal S, Mulet A. Influence of temperature and ultrasound on drying kinetics and antioxidant properties of red pepper. Drying Technology. 2019;**37**:1-8. DOI: 10.1080/07373937.2018.1473417

[71] Musielak G, Mierzwa D, Kroehnke J. Food drying enhancement by ultrasound. A review. Trends in Food Science and Technology. 2016;**56**:126-141. DOI: 10.1016/j.tifs.2016.08.003

[72] De la Fuente-Blanco S et al. Food drying process by power ultrasound. Ultrasonics. 2006;**44**:523-527. DOI: 10.1016/j.ultras.2006.05.181

[73] Yao Y. Enhancement of mass transfer by ultrasound: Application to adsorbent regeneration and food drying/dehydration. Ultrasonics Sonochemistry. 2016;**31**:512-531. DOI: 10.1016/j.ultsonch.2016.01.039

[74] Cárcel JA, Benedito J, Rosselló C, Mulet A. Influence of ultrasound intensity on mass transfer in apple immersed in a sucrose solution. Journal of Food Engineering. 2007;**78**:472-479. DOI: 10.1016/j.jfoodeng.2005.10.018

[75] Gallego-Juárez JA, Rodríguez G, Acosta V, Riera E. Power ultrasonic transducers with extensive radiators for industrial processing. Ultrasonics

Sonochemistry. 2010;**17**:953-964. DOI: 10.1016/j.ultsonch.2009.11.006

[76] García-Pérez JV et al. Ultrasonic drying of foodstuff in a fluidized bed: Parametric study. Ultrasonics. 2006;**44**:539-543. DOI: 10.1016/j.ultras.2006.06.059

[77] Garcia-Perez JV, Rossello C, Carcel JA, De la Fuente S, Mulet A. Effect of air temperature on convective drying assisted by high power ultrasound. Defect and Diffusion Forum. 2006;**258**:563. DOI: 10.4028/www.scientific.net/DDF.258-260.563

[78] Tekin Z, Başlar M, Karasu S, Kilicli M. Dehydration of green beans using ultrasound-assisted vacuum drying as a novel technique: Drying kinetics and quality parameters. Journal of Food Processing & Preservation. 2017;**41**:e13227. DOI: 10.1111/jfpp.13227

[79] Chen ZG, Guo XY, Wu T. A novel dehydration technique for carrot slices implementing ultrasound and vacuum drying methods. Ultrasonics Sonochemistry. 2016;**30**:28-34. DOI: 10.1016/j.ultsonch.2015.11.026

[80] Cheng XF, Zhang M, Adhikari B. Effect of ultrasonically induced nucleation on the drying kinetics and physical properties of freeze-dried strawberry. Drying Technology. 2014;**32**:1857-1864. DOI: 10.1080/07373937.2014.952741

[81] Schössler K, Jäger H, Knorr D. Novel contact ultrasound system for the accelerated freeze-drying of vegetables. Innovative Food Science & Emerging Technologies. 2012;**16**:113-120. DOI: 10.1016/j.ifset.2012.05.010

[82] Cárcel JA et al. High intensity ultrasound effects on meat brining. Meat Science. 2007;**76**:611-619. DOI: 10.1016/j.meatsci.2007.01.022

[83] Legay M, Gondrexon N, Le Person S, Boldo P, Bontemps A. Enhancement of heat transfer by ultrasound: Review and recent advances. International Journal of Chemical Engineering. 2011;**2011**:670108. DOI: 10.1155/2011/670108

[84] Gondrexon N, Rousselet Y, Legay M, Boldo P, Le Person S, Bontemps A. Intensification of heat transfer process: Improvement of shell-and-tube heat exchanger performances by means of ultrasound. Chemical Engineering and Processing Process Intensification. 2010;**49**:936-942. DOI: 10.1016/j.cep.2010.06.007

[85] Uhlenwinkel V, Meng RX, Bauckhage K. Investigation of heat transfer from circular cylinders in high power 10 kHz and 20 kHz acoustic resonant fields. International Journal of Thermal Sciences. 2000;**39**:771-779. DOI: 10.1016/S1290-0729(00)00270-2

[86] Pohlman FW, Dikeman ME, Kropf DH. Effects of high intensity ultrasound treatment, storage time and cooking method on shear, sensory, instrumental color and cooking properties of packaged and unpackaged beef pectoralis muscle. Meat Science. 1997;**46**:89-100

[87] Cichoski AJ et al. Ultrasound-assisted post-packaging pasteurization of sausages. Innovative Food Science & Emerging Technologies. 2015;**30**:132-137. DOI: 10.1016/j.ifset.2015.04.011

[88] Condón-Abanto S et al. An assessment of the application of ultrasound in the processing of ready-to-eat whole brown crab (*Cancer pagurus*). Ultrasonics Sonochemistry. 2018;**40**(Part A):497-504. DOI: 10.1016/j.ultsonch.2017.07.044

[89] Wang Y, Zhang W, Zhou GH. Effects of ultrasound-assisted frying on the physiochemical properties and microstructure of fried meatballs.

International Journal of Food Science and Technology. 2019;**54**:2915-2926. DOI: 10.1111/ijfs.14159

[90] Silva J et al. Is it possible to reduce the cooking time of mortadellas using ultrasound without affecting their oxidative and microbiological quality? Meat Science. 2020;**159**:107947. DOI: 10.1016/j.meatsci.2019.107947

[91] Álvarez I et al. Cooking device. 2014. Patent EP2840866B1

[92] Cidad-Hidlago S. Aplicación de ultrasonidos en el cocinado de alimentos [thesis]. Zaragoza (Spain): Universidad de Zaragoza; 2018

[93] Gaukel V. Cooling and freezing of foods. Reference Module in Food Science. Elsevier; 2016. ISBN 9780081005965. DOI: 10.1016/B978-0-08-100596-5.03415-6

[94] Zheng L, Sun D-W. Innovative applications of power ultrasound during food freezing processes—A review. Trends in Food Science and Technology. 2006;**17**:16-23. DOI: 10.1016/j.tifs.2005.08.010

[95] Hu SQ, Liu G, Li L, Li ZX, Hou Y. An improvement in the immersion freezing process for frozen dough via ultrasound irradiation. Journal of Food Engineering. 2013;**114**:22-28. DOI: 10.1016/j.jfoodeng.2012.07.033

[96] Li B, Sun DW. Novel methods for rapid freezing and thawing of foods: A review. Journal of Food Engineering. 2002;**54**:175-182. DOI: 10.1016/S0260- 8774(01)00209-6

[97] Xin Y et al. The effects of ultrasound-assisted freezing on the freezing time and quality of broccoli (*Brassica oleracea* L. var. *botrytis* L.) during immersion freezing. International Journal of Refrigeration. 2014;**41**:82-91. DOI: 10.1016/j.ijrefrig.2013.12.016

[98] Delgado AE, Zheng LY, Sun DW. Influence of ultrasound on freezing rate of immersion-frozen apples. Food and Bioprocess Technology. 2009;**2**:263-270. DOI: 10.1007/s11947-008-0111-9

[99] Islam MN, Zhang M, Adhikari B, Cheng XF, Xu BG. The effect of ultrasound- assisted immersion freezing on selected physicochemical properties of mushrooms. International Journal of Refrigeration. 2014;**42**:121-133. DOI: 10.1016/j.ijrefrig.2014.02.012

[100] Zhang M, Niu H, Chen Q, Xia X, Kone B. Influence of ultrasound-assisted immersion freezing on the freezing rate and quality of porcine longissimus muscles. Meat Science. 2018;**136**:1-8. DOI: 10.1016/j.meatsci.2017.10.005

[101] Sun Q et al. Ultrasound-assisted immersion freezing accelerates the freezing process and improves the quality of common carp (*Cyprinus carpio*) at different power levels. LWT- Food Science and Technology. 2019;**108**:106-112. DOI: 10.1016/j.lwt.2019.03.042

[102] Miles CA, Morley MJ, Rendell M. High power ultrasonic thawing of frozen foods. Journal of Food Engineering. 1999;**39**:151-159

[103] Gambuteanu C, Alexe P. Comparison of thawing assisted by low-intensity ultrasound on technological properties of pork *Longissimus dorsi* muscle. Journal of Food Science and Technology. 2015;**52**:2130-2138. DOI: 10.1007/s13197-013-1204-7

[104] Li X, Sun P, Ma Y, Cai L, Li J. Effect of ultrasonic thawing on the water holding capacity, physico-chemical properties, and structure of frozen tuna fish (*Thunnus tonggol*) myofibrillar proteins. Journal of the Science of Food and Agriculture. 2019;**99**:5083-5091. DOI: 10.1002/jsfa.9752

[105] Rodríguez G et al. Experimental study of defoaming by air-borne ultrasonic technology. Physics Procedia. 2010;**3**:135-139. DOI: 10.1016/j.phpro.2010.01.019

[106] Paniwnyk L. Other non-thermal processing technics: Application of ultrasound. In: Sun DW, editor. Emerging Technologies for Food Processing. London: Academic Press; 2014. pp. 271-288

[107] De-Sarabia ERF, Gallego-Juárez JA, Mason TJ. Airborne ultrasound for the precipitation of smokes and powders and the destruction of foams. Ultrasonics Sonochemistry. 2006;**13**:107-116. DOI: 10.1016/j.ultsonch.2005.04.001

[108] Villamiel M, Verdurmen R, de Jong P. Degassing of milk by high- intensity ultrasound. Milchwissenschaft. 2000;**55**:123-125

[109] Mc Clements DJ. Advances in the application of ultrasound in food analysis and processing. Trends in Food Science and Technology. 1995;**6**:293-299

[110] Alarcon-Rojo AD, Janacua H, Rodriguez JC, Paniwnyk L, Mason TJ. Power ultrasound in meat processing. Meat Science. 2015;**107**:83-96. DOI: 10.1016/j.meatsci.2015.04.015

[111] Lyng JG, Allen P, Mckenna BM. The influence of high intensity ultrasound baths on aspects of beef tenderness. Journal of Muscle Foods. 1997;**8**:237-249

[112] Jayasooriya SD, Bhandari BR, Torley P, D'Arey BR. Effect of high power ul- trasound waves on properties of meat: A review. International Journal of Food Properties. 2004;**7**:301-319. DOI: 10.1081/JFP-120030039

[113] Bhat ZF, Morton JD, Mason SL, Bekhit AEDA. Applied and emerging methods for meat tenderization: A comparative perspective. Comprehensive Reviews in Food Science and Food Safety. 2018;**17**:841-859. DOI: 10.1111/1541-4337.12356

[114] Available from:https://www.hielscher.com/ultrasonic-meat-tenderization.htm [Accessed: October 2019]

[115] Rawson FF. An introduction to ultrasonic food cutting. In: Povey MJW, Mason TJ, editors. Ultrasound in Food Processing. London: Blackie Academic and Professional; 1998. pp. 254-269

Chapter 8

Application of Ultrasonics on Preparation of Magnesium Alloys

Xingrui Chen and Qichi Le

Abstract

This chapter mainly describes the application of ultrasonic on preparation of magnesium alloys, which includes all of the interesting and novel research results of authors in the past decade. The chapter includes the following topics to readers: the ultrasonic cavitation behavior in magnesium alloys; the effect of ultrasonic treatment on melt structure; ultrasonic degassing of magnesium alloys; effect of ultrasonic melt treatment on microstructure of magnesium alloys; dual-frequency ultrasonic treatment on solidification of magnesium alloys; and ultrasonic direct-chill (DC) casting process of magnesium alloys billets, including the novel variable-frequency ultrasonic technology. The chapter almost covers all the aspects of ultrasonic application on preparation of magnesium alloys and can help readers have a systematic understanding of ultrasonic melt treatment on magnesium alloys.

Keywords: ultrasonics, magnesium alloy, melt treatment, casting, degassing

1. Introduction

Magnesium alloy is the lightest metal structure material until now. It has caught a lot of attentions from industrial manufacturers all over the world. It is used in many fields such as vehicles, aircrafts, and military industry, due to its great advantages including low density, high strength/weight ratio, etc. [1, 2], while the unique crystal structure, the existence of microporosity, and coarse microstructure limit its application and development. To solve these problems, researchers have developed some technologies. For example, the semisolid process technology [3] and rapid solidification technology [4] can refine the Mg grain, and the argon degassing technology can remove the hydrogen from the melt. However, the ultrasonic melt treatment technology is a best choice considering its environmental friendly and easy realized characteristics.

After decades of development, this technology has been employed in many aspects for preparation of magnesium alloy. This chapter is aimed at introducing the role of ultrasound on metallurgy process of magnesium alloys. We hope people who are interested in this field may be inspired by this chapter.

2. Ultrasonic cavitation in magnesium alloy

The reason why ultrasonic melt treatment technology attracts many researchers is that it has excellent ability to refine the microstructure of alloys. This improvement is identified as the contribution of ultrasonic cavitation. In other words, the cavitation holes the key to ultrasonic refinement. Therefore, it is of great importance to investigate the cavitation behavior in magnesium alloy. However, because of the fact that the characteristics of magnesium melt such as high temperature and opacity, the numerical simulation is a viable method to reflect the cavitation behavior.

The cavitation behavior in magnesium melt is calculated the Rayleigh–Plesset Equation [5]:

$$R\left(\frac{d^2R}{dt^2}\right)+\frac{3}{2}\left(\frac{dR}{dt}\right)^2=\frac{1}{\rho}\left[\left(P_0+\frac{2\sigma}{R_0}\right)\left(\frac{R_0}{R}\right)^{3\kappa}-\frac{2\sigma}{R}-\frac{4\mu}{R}\left(\frac{dR}{dt}\right)-P_0+P_v+p\right] \tag{1}$$

where R is the radius of cavitation bubble, R_0 is the initial radius of bubble, P_0 is the initial pressure, P_v is the vapor pressure, p is the external acoustic pressure, σ is the surface tension, μ is the fluid viscosity, and κ is the polytropic exponent. This simulation does not consider the variation of the ultrasonic intensity and the liquid temperature. The gas and vapor are incompressible. The motion of cavitation bubble wall is set as spherical symmetric movement.

To calculate the cavitation behavior under single-frequency ultrasonic vibration, the p in Eq. (1) is defined as $p = P_m sin(2\pi f)$, where P_m is the acoustic pressure amplitude and f is the frequency of ultrasound. **Table 1** lists the parameters used in this calculation.

Figure 1 shows the effect of ultrasonic frequency on cavitation bubble's behavior in magnesium alloy. It notes that frequency has strong effect on bubble's behavior. Ultrasound with lower frequency generally has larger-sized cavitation bubbles. The 22 kHz is an important frequency, which can be defined as the threshold frequency of instantaneous cavitation for magnesium alloy. When the frequency is larger than 22 kHz, bubbles usually suffer several shrinkages and expansions before collapse. It also notes that ultrasound with higher frequency is not beneficial to the melt treatment of magnesium alloy. The initial radius of bubble is another key to cavitation. **Figure 2** shows the effect of bubble's radius on cavitation behavior of magnesium alloy. Small initial bubble changes regularly under single-frequency ultrasonic vibration. When bubble's radius is close to resonance frequency, cavitation bubble's radius changes rapidly in a short time. It should be pointed out that the initial bubble in magnesium melt is provided by hydrogen bubble, suggesting that they cannot be controlled.

Name	Value	Description
R_0	0.1–100 μm	Initial radius of cavitation bubble
P_0	$1.013{\cdot}10^5$ Pa	Initial pressure outside cavitation bubble
P_v	1000 Pa	Vapor pressure within cavitation bubble
σ	0.564	Surface tension
μ	$1.12{\cdot}10^{-3}$	Fluid viscosity
κ	1.34	Polytropic exponent
f	20 KHz	Initial frequency of another wave
P_m	$1.01{\cdot}10^6$ Pa	Acoustic pressure amplitude

Table 1.
Parameters and initial values for Eq. (1).

When dual-frequency ultrasonic field (DUF) is introduced to magnesium melt, the cavitation behavior becomes completely different, as shown in **Figure 3**. Small bubble's radius becomes irregular. It is worth to note that bubble becomes larger under dual-frequency ultrasonic field. The bubble cannot collapse within 250 μs when the initial radius is over 15 μm under single-frequency ultrasonic field, but

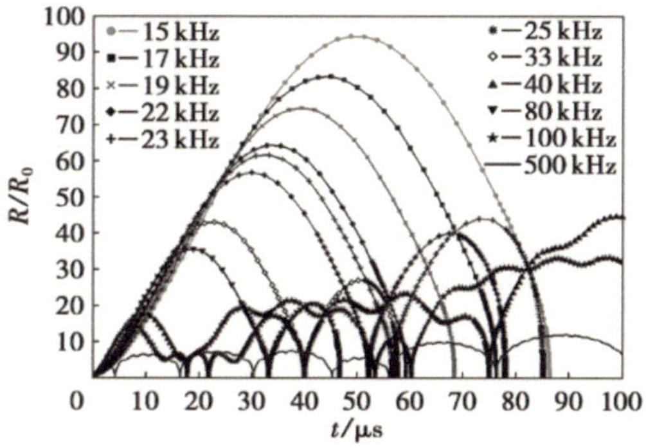

Figure 1.
Cavitation bubbling behavior with different ultrasonic frequencies.

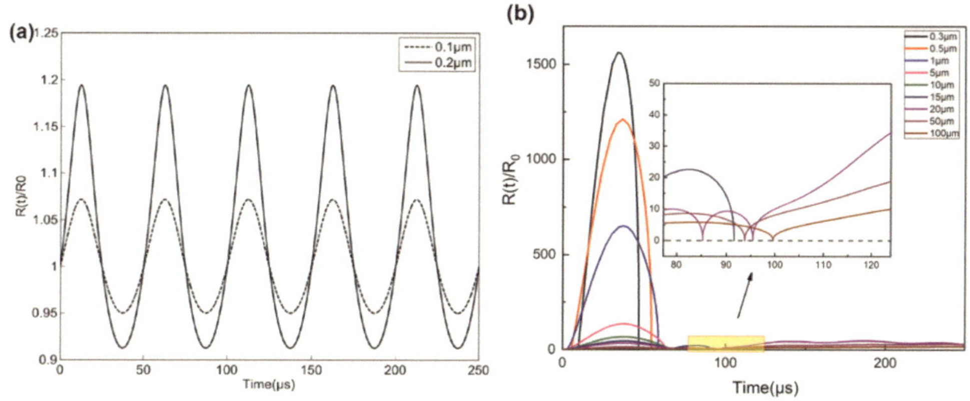

Figure 2.
Cavitation bubbling behavior with different initial bubble radii under single-frequency ultrasonic field: (a) 0.1 μm and 0.2 μm; (b) 0.3-100 μm.

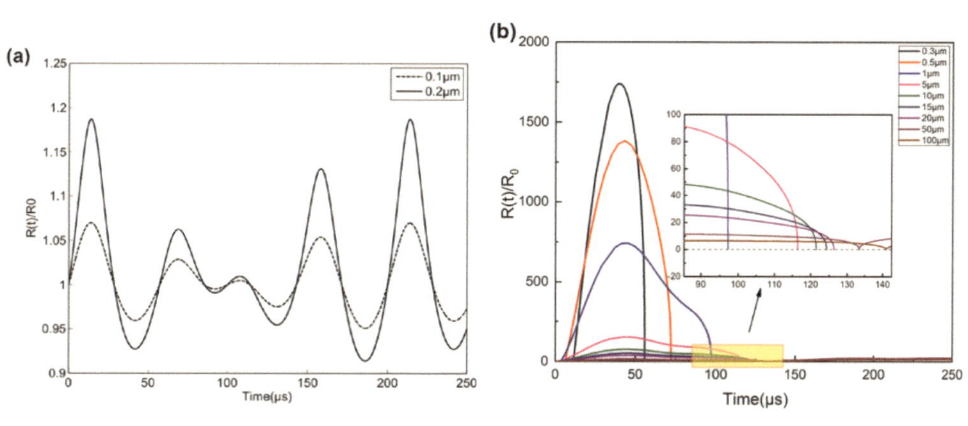

Figure 3.
Cavitation bubbling behavior with different initial bubble radii under dual-frequency ultrasonic field: (a) 0.1 μm and 0.2 μm; (b) 0.3-100 μm.

this limitation is expanded to 20 μm under dual-frequency ultrasonic field. Thus, the dual-frequency ultrasonic field has great potential to enhance the cavitation in magnesium alloy, and the experimental results are shown in Section 6 of this chapter.

3. Effect of ultrasonic treatment on melt structure

The ultrasonic vibration is a mechanical wave. When the wave propagates in the magnesium melt, molecules suffer regular vibration, which can change the melt structure. To investigate this issue, we employ the electrical resistivity to character the melt structure. The experimental setup is shown in **Figure 4**. The four-electrode method is used. The tested melt is Pb–Sn liquid metal. The electrical resistivity ρ is calculated by

$$\rho = [R - (R_e + R_w + R_c)]S/L \quad (2)$$

where R is the total electrical resistance, Re is the resistance of electrodes, R_w is the resistance of wire, R_c is the contact resistance, S is the cross area of melt, and L is the length of melt. In order to avoid the interference of ultrasonic horn during measurement, the horn is coated by ZrO_2–CaO.

Figure 5 displays the variation of electrical resistivity for the Pb–80 wt.%Sn melt at 553 K and the relevant microstructure evolution. Four stages for the electrical resistivity ρ are observed. The resistance keeps stable at ρ_0 before ultrasonic vibration. The resistance reduces to ρ' immediately when the ultrasound is introduced to the melt. The drop is defined as $\Delta\rho$. The resistance rises to ρ'' instantaneously once the ultrasound disappeared. In addition, the $\Delta\rho'$ is defined to describe the difference between ρ_0 and ρ. It is also found that the ρ stays at ρ'' for Δt_1 ($\Delta\rho' = \rho_0 - \rho''$). After the holding time Δt_2, the ρ returns to the initial value (ρ_0). The microstructure evolution is shown in **Figure 5** as well. Without ultrasonic treatment, developed dendrites and large columnar crystals are observed with some branch length of ~400 μm, as shown in **Figure 5(a)**. Microstructure is dramatically refined with ultrasonic vibration 5 min after treatment. With the increase of holding time, the microstructure also recovers to the initial status gradually.

In this work, the modified band theory is employed to understand the electrical properties of melt. Thus, the dimensionless parameter δ is introduced to show the disorder degree of electrons:

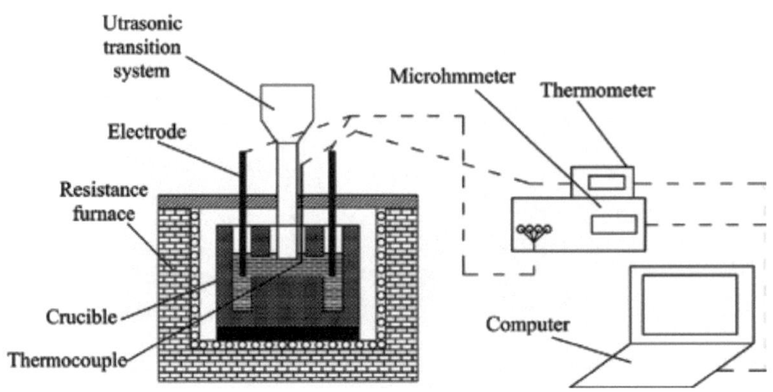

Figure 4.
Schematic of the experimental setup for electrical resistivity tests.

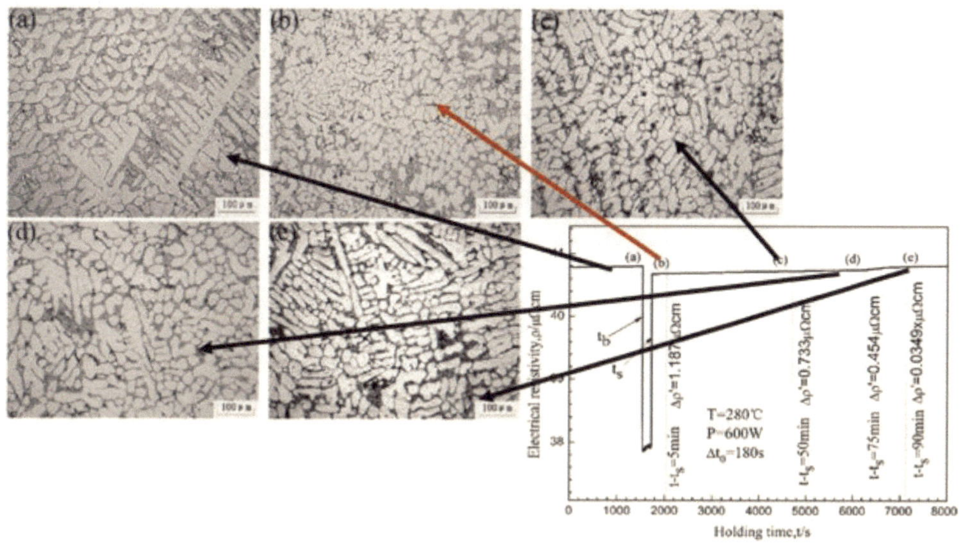

Figure 5.
Electrical resistivity ρ vs. the holding time t for the Pb–80 wt.%Sn melt at 553 K and microstructure evolution.

$$\delta = W/B \tag{3}$$

where W and B represent the width of electron energy' distribution and energy band, respectively. There is a strong relativity between the degrees of system disorder and localization of electron [6]. The number of localized electrons increases with the rise of disorder degree of the system. In addition, the localized electrons are adverse to the electrical conductivity, while the extended electrons are beneficial to the conductivity. Therefore, the electrical conductivity (σ) associated with the energy band theory is shown as follows:

$$\sigma = ne^2\tau(E_F)/m^* \tag{4}$$

where n represents the number density of extended electrons in the investigated liquid, τ represents the electron's free time, m^* represents the effective mass, and E_F is the Fermi energy. Thus, the number density of extended electrons holds the key to the electrical conductivity of melt, based on Eq. (4).

When ultrasound is introduced into the Pb–Sn melt, the atoms are forced to vibrate due to the characteristics of sound wave. Then the disordered atoms become "well-organized" under ultrasonic field, which means that the disorder degree of the melt is reduced. Thus, with the decrease of disorder degree, the W is reduced, and consequently, the number density of extended electrons is increased, resulting in the improvement of electrical conductivity.

The forced vibration disappears when the ultrasonic vibration is stopped. "Well-organized" atoms recover their disorder degree, increasing the number density of localized electron. It is worth to notice that a long time is required to recover the initial value of resistivity, suggesting that the ultrasonic treatment can change the short-ordered structures (SROs) of melt [7].

4. Ultrasonic degassing of magnesium alloy

Degassing is always a significant process during the melting of magnesium alloys [8–10]. At the end of solidification process, micropores are formed because of the pressure decrease of interdendritic flow or existence of dissolved gaseous elements in the liquid alloy. To degas and purify the melt, researchers have developed some technologies such as gaseous chlorine, hexachlorethane (C_2Cl_6), rotating impeller degasser with argon, and ultrasonic degassing. Among these methods, the ultrasonic degassing is the most convenient way without polluting the melt. The flux and argon are employed to help ultrasonic degassing reach the best efficiency. **Figure 6** shows the experimental equipment. A channel is set inside the ultrasonic horn as the path for argon and flux powders.

Figure 7 shows the hydrogen content with the increase of holding time by different degassing technologies. The initial value of hydrogen content is 22.5 ppm, which reduces dramatically after degassing process. It notes that the ultrasonic argon degassing technology has the highest efficiency, displaying the lowest level of hydrogen content in the same time. It is interesting to note that the degassing efficiency of ultrasonic degassing technology is lower than the argon degassing technology. The hollow ultrasonic horn should be responsible for this drop, which reduce the propagation efficiency of ultrasound.

For magnesium alloy, the hydrogen is removed by the tiny and dispersed argon gas bubbles. The hydrogen starts accumulating in argon gas due to the balance with

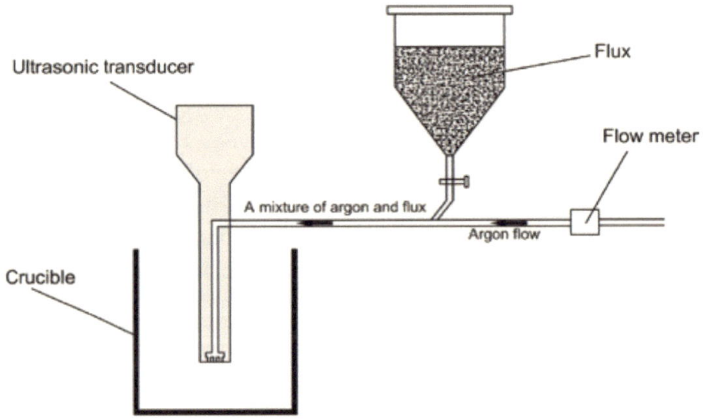

Figure 6.
The schematic of the experimental equipment for ultrasonic degassing.

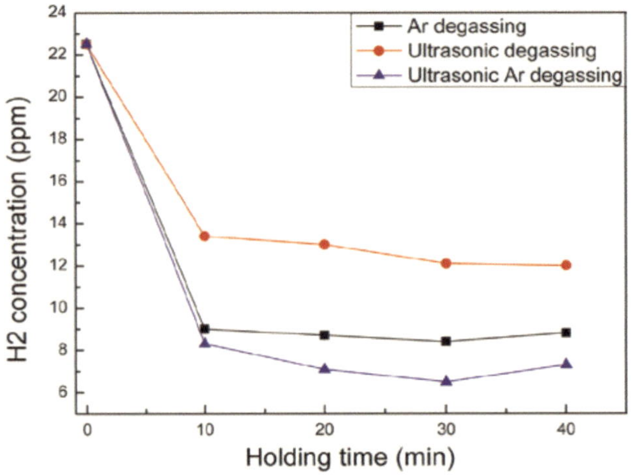

Figure 7.
Hydrogen content as a function of holding time in different degassing processes.

the dissolved concentration in the melt. The hydrogen transports from the melt to the bubble boundary firstly. The hydrogen reacts and become the hydrogen gas. Finally, the hydrogen is transported away as the gas forms.

According to the results above, the hydrogen concentration has functional relation with time of purification t [11]:

$$[\mathrm{H}]/[\mathrm{H}]_0 = \exp[-(k\rho S/M) \cdot t] \tag{5}$$

where k is the mass transfer coefficient; ρ is the density of magnesium melt; S is the bubble's surface area; $[\mathrm{H}]_0$ is the initial hydrogen content; and M is the weight of magnesium melt.

Obviously, the hydrogen content deceases with the increase of degassing time while the point is improving the degassing efficiency. Based on Eq. (5), the degassing efficiency is determined by the mass transfer coefficient k and the total surface area of the bubbles S in the same time. Thus, to promote the degassing efficiency, increasing the k and S is feasible. The bubble's surface area is driven by

$$S = 3\dot{G}h/vr = 3\dot{G}\tau/r \tag{6}$$

where h is the bath height; v is the velocity of bubble; r is the radius of bubble; and τ is the residual time in the melt. The velocity of bubble has a strongly positive relation with its radius [12], suggesting that larger bubble usually has higher velocity. However, the mass transfer coefficient k has negative relation with the bubble's diameter, implying small bubble has higher k value. Thus, to raise the degassing efficiency, employing the small sized bubble is critical.

The cavitation and rectified mass diffusion are two mechanisms for ultrasonic degassing process. Cavitation bubble is formed when the acoustic pressure surpasses the cavitation threshold. Generally, there are two nucleating mechanisms for cavitation, namely, the microscopy gas cavities and non-wettable particles. In case of magnesium melt, both of two mechanisms are applicable, due to the large content of oxides and high level of hydrogen solubility. The cavitation bubble starts expanding and shrinking under the alternating pressure. The dissolved hydrogen can diffuse into the cavitation bubble during the shrinkage process, so-called rectified mass diffusion. In this way, hydrogen is accumulated to form large gas bubble which can escape form the melt.

It is noticed that the argon-ultrasonic degassing has best efficiency. Interaction between argon and ultrasound should respond to this result. Firstly, the argon gas is purified through the center of ultrasonic horn. The purified argon bubbles start expanding and shrinking because of the propagation of ultrasound. These bubbles can be broken into dense small bubbles under high-intensity acoustic pressure, increasing the surface area of bubble. Besides, the rectified mass diffusion effect brought by ultrasonic vibration can encourage hydrogen diffusing to the argon bubbles. In addition, the acoustic streaming can carry the refined argon bubble to the whole melt. What is more, the argon bubbles can also improve the forming of cavitation and make bubbles easily to ascend to the melt surface. Therefore, these mechanisms work together and promote the degassing efficiency.

5. Effect of ultrasonic melt treatment on microstructure of magnesium alloy

Magnesium alloys have a unique structure, namely, the HCP structure, which has limited slip system. In addition, magnesium alloy has a large Hall–Petch slope, k, according to the relationship $\sigma = \sigma_0 + kd^{-1/2}$. These two characteristics make the grain refinement become one of the important methods to improve the mechanical properties and the workability of magnesium alloy. Ultrasonic melt treatment has proved its ability to control and refine the solidification structure for many alloys. We investigated the ultrasonic treatment on microstructure evolution of magnesium alloys [13, 14].

The experimental setup for this study is shown in **Figure 8**. The treatment system includes a resistance furnace, an iron crucible, a water tank, and an ultrasonic vibration system with the maximum power of 2000 W. The ultrasonic vibration system is composed by an ultrasonic generator, a magnetostrictive transducer, and a mild steel-made acoustic horn.

One of the widely used commercial magnesium alloys, AZ80 alloy, was chosen in this work. Its chemical composition (wt.%) was as follows: Al 8.0, Zn 0.6, Mn 0.15, and Mg balance. The prepared materials were melted in the crucible with heating by the resistance furnace and protection of CO2 + 0.5% SF6 atmosphere. The temperature of melt maintained at 650°C for 600 s. The preheated ultrasonic horn (650°C) was inserted 0.02 m under the surface of melt. After treatment, the melt was quenched in the water tank immediately.

The microstructures of as-cast AZ80 alloy without and with ultrasonic treatment are shown in **Figure 9(a)** and **(b)**, respectively. Coarse dendrites and large-sized

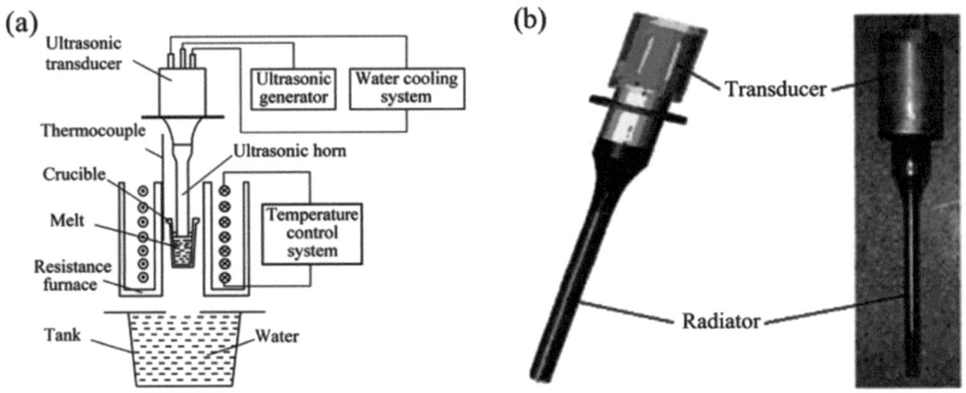

Figure 8.
(a) Schematic image of experimental apparatus and (b) structures of ultrasonic transducer and ultrasonic radiator.

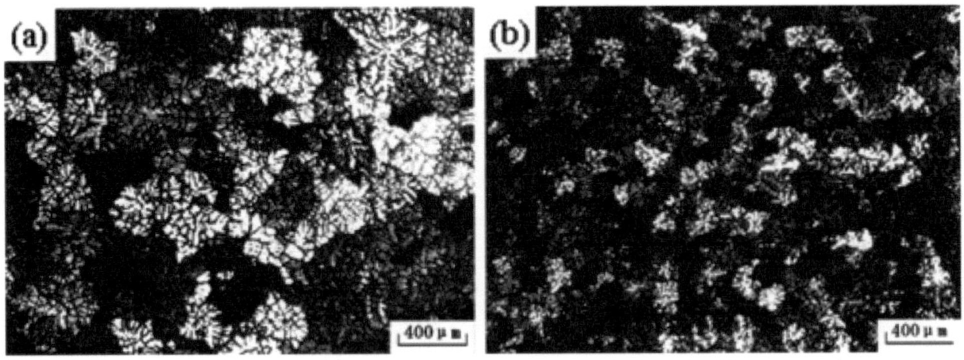

Figure 9.
Microstructure of the AZ80 alloy casted (a) without ultrasonic and (b) with 600 W ultrasonic treatment (φ0.02 m ultrasonic emitter used).

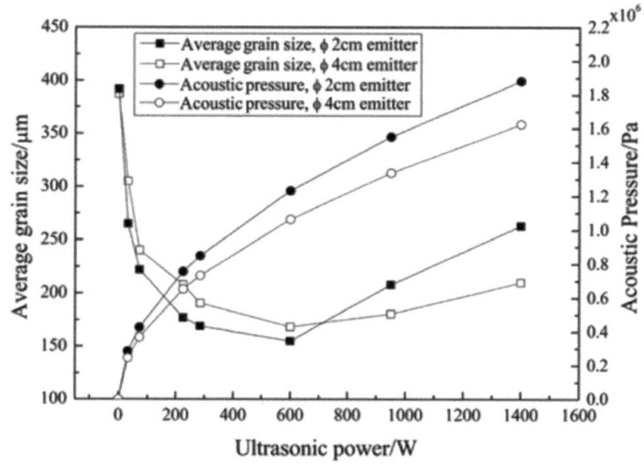

Figure 10.
Influence of ultrasonic power on the grain size of the AZ80 ingot and the relevant acoustic pressure associated with treated sample (solid mark, φ0.02 m ultrasonic emitter used; hollow mark, φ0.04 m ultrasonic emitter used).

primary grains are observed with the grain size of about 387 μm, suggesting the typical dendrite growth mode. With ultrasonic treatment, the grain size is rapidly reduced to 147 μm with the observation of globular and even primary grains. In addition, the homogeneity is improved as well. **Figure 10** shows the effect of ultrasonic power on grain refinement efficiency of AZ80 alloy in the same treatment time and the relation between acoustic pressure and ultrasonic power. When the ultrasonic power rises from 0 to 600 W, the acoustic pressure increases, but the grain size shows adverse trend, suggesting better grain refinement efficiency. With the further increment of ultrasonic power, the acoustic pressure continues to rise, while grain size stops reducing. In this case, the best grain refinement efficiency appears at 600 W.

The grain refinement of AZ80 magnesium alloy of the ultrasonic treatment can be attributed to the ultrasonic cavitation. The ultrasonic cavitation can bring two influences which can affect the nucleation and growth of grains, so-called cavitation-enhanced nucleation and cavitation-induced (shock waves) dendrite fragmentation [15, 16]. In this work, the treatment temperature is 40°C higher than the liquidus temperature of treated alloy, which means that the cavitation-induced dendrite fragmentation does not work and has no contribution on the grain refinement. Therefore, the cavitation-enhanced nucleation is the main reason that caused the grain refinement. The ultrasonic cavitation can clean the surfaces of the poorly wetted particles in the melt, and consequently, these particles become the nucleation points, increasing the number of primary magnesium grains. With the help of acoustic streaming, these effective nucleation sites are carried to the whole melt. Note the ultrasonic cavitation works when ultrasonic pressure exceeds the cavitation threshold. Thus, strong acoustic pressure can enlarge and enhance the cavitation effect. With the increase of ultrasonic power, the acoustic pressure is increased, which brings better efficiency of grain refinement. However, the results reflect that higher ultrasonic power, e.g., 950 W or even 1400 W, cannot further enhance the grain refinement efficiency but weaken it. When the cavitation bubble collapses, high temperature and pressure are released in melt, so-called thermal effect. The higher ultrasonic power may also increase the thermal effect. Therefore, the cooling rate is reduced, which is beneficial to the growth of grain but has adverse effect on grain refinement caused by ultrasonic treatment.

6. Dual-frequency ultrasonic treatment on solidification behavior of magnesium alloy

In Section 1, we discussed about the effect of dual-frequency ultrasound on cavitation bubble's behavior in magnesium melt. In this section, we will show the experimental results. In recent decades, dual-frequency acoustic vibration has been employed in sonochemistry, medical diagnostics, and fluid engineering. For example, the dual-frequency ultrasonic wave can enhance the extraction rates of leaching process and improve the enzymolysis of corn gluten meal [17]. The dual-frequency ultrasonic vibration mode is also used to detect the fluid pressure and the bubble density in liquid [18, 19]. Also, dual-frequency ultrasound can help increase the accuracy of biomedical diagnosis. We firstly employed the dual-frequency ultrasonic field (DUF) to treat the magnesium alloy and investigated its solidification behavior [20].

Figure 11 shows the experimental apparatus used in this work. Two self-governed ultrasonic systems were used simultaneously. The frequency of two systems is 15 and 20 kHz, respectively. Each ultrasonic vibration system is made up by an ultrasonic power supply, an ultrasonic transducer, an acoustical waveguide, and a Ti6Al4V acoustic horn. A vortex tube cooler was used to ensure the comfortable working environment of transducer. The commercial ZK60 was chosen as the treated magnesium alloy.

Figure 12(a)–(f) displays the optical images of as-cast ZK60 alloys with and without different ultrasonic treatments. Without ultrasonic treatment, developed dendrites and coarse grains are observed with large size. Similar with the results in Section 5, the SUF can refine the α-Mg grains in some degree with the disappearance of dendrites. The DUF can considerably refine the α-Mg grains, and the refinement efficiency is different according to electric power. The morphology of α-Mg grains also changes a lot, showing the fine and globular form. The grain size of

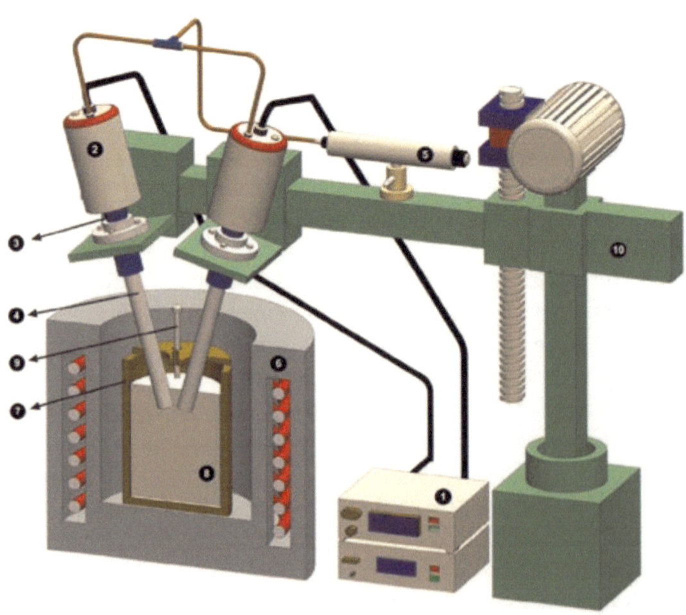

Figure 11.
Schematic image of the experimental apparatus: (1) ultrasonic power supply, (2) ultrasonic transducer, (3) acoustical waveguide, (4) acoustic radiator/horn, (5) vortex tube cooler, (6) resistance furnace, (7) iron crucible, (8) magnesium melt, (9) thermocouple, (10) positioning device.

untreated and SUF-treated ZK60 alloys is 183 ± 8 μm and 125 ± 6 μm, respectively. In the same total electric power, the DUF makes the grain size reduce to 72 ± 6 μm. Increasing the electric power, the grain size reduces to 59 ± 4 μm (1000 W) and 50 ± 3 μm (1400 W), while further increment of electric power cannot decrease the grain size but rise to 58 ± 3 μm after 1800 W DUF treatment, which has the same results in Section 5.

Now it is clear that the DUF has the higher refinement efficiency than SUF in the same total electric power. To better explain the mechanism of ultrasonic treatment on grain refinement of the investigated alloy, a schematic image is given based on the microstructure observation, as shown in **Figure 13**. The ZK60 melt suffered cyclic alternating acoustic field, when the ultrasonic vibration is introduced. The liquid molecules are pulled to form the cavitation bubbles in negative acoustic period. The enlargement and formation of cavitation bubble can absorb heat from the melt, resulting in the local supercooling on the surface of bubbles, as shown in **Figure 13(a)**. Consequently, nucleation can easily occur on bubble's surface,

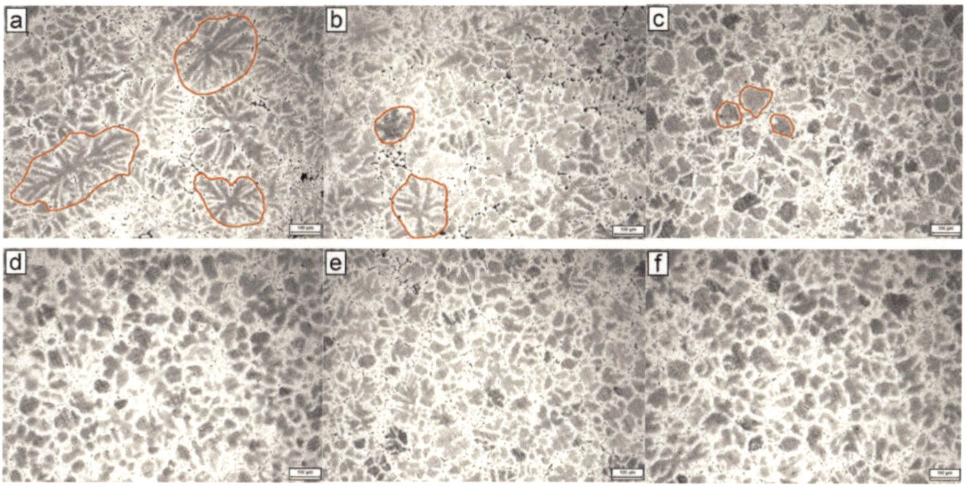

Figure 12.
Optical images of as-cast ZK60 alloys with different ultrasonic treatments: (a) without ultrasonic treatment; (b) SUF treatment with 1200 W; and DUF treatment with (c) 600 W; (d) 1000 W; (e) 1400 W; and (f) 1800 W electric power.

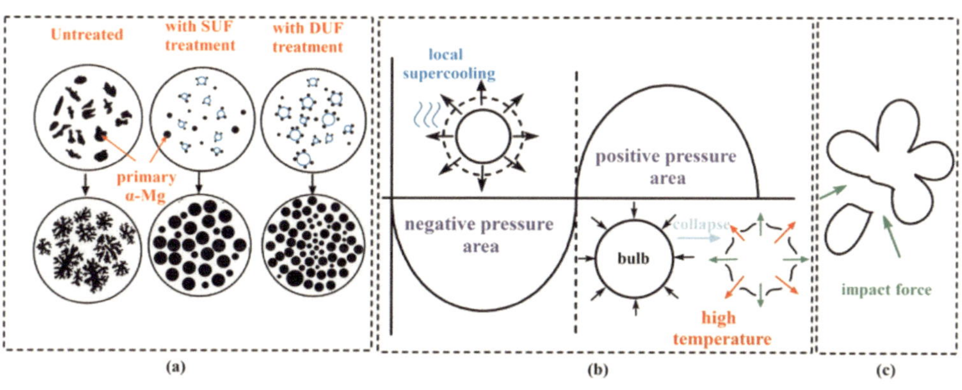

Figure 13.
Schematic image of mechanisms of ultrasonic grain refinement: (a) cavitation-enhanced heterogeneous nucleation; (b) formation, growth, and collapse of the cavitation bubble; and (c) dendrite fragmentation caused by impact force.

increasing the number of primary nuclei. In the positive acoustic pressure period, the cavitation bubble collapsed, releasing the high temperature and strong impact force (seen in **Figure 13(b)**). This impact force fragments dendrites and primary Mg nuclei (as shown in **Figure 13(c)**), by which the number of nuclei is also raised. Finally, the acoustic streaming, a liquid flow due to acoustic pressure gradient, transports the increased nuclei through the whole magnesium alloy melt.

In Section 1, the numerical results show that the DUF can produce larger cavitation bubbles and increase the number of instantaneous bubbles. Thus, these two improvements enhance the grain refinement efficiency. Larger cavitation bubble can absorb more heat from melt during its growth process, by which the heterogeneous nucleation caused by ultrasonic treatment is improved (as shown in **Figure 13(a)**). In addition, stronger impact force is released by larger cavitation bubble, which enhances the fragmentation of primary grains.

Besides α-Mg grain, the ultrasonic treatment also has ability to change the morphology of β-phase in ZK60, as shown in **Figure 14**. A large number of lamellar β-phases are observed in untreated sample. There are few dot-like β-phase in the α-Mg matrix. The SUF makes some lamellar β-phase become the strip-like form. The number of dot-like β-phase increases. A lot of tiny dot-like β-phases with dispersed distribution can be observed after DUF treatment. The energy-dispersive spectroscopy (EDS) results reflect these three aerial Mg-Zn phases.

To further confirm the type of these phases, XRD tests are carried out, as shown in **Figure 15**. Thus, in combination with the XRD patterns and EDS results (**Table 2**), three β-phases observed from SEM images are laves phases ($MgZn_2$). It also notes that the ultrasonic treatment cannot change the phase constitutions magnesium alloy, while it is interesting to find that the orientation of α-Mg is changed. According to partial enlarged drawing of **Figure 15**, the ultrasonic treatment reduces the intensities of diffraction peaks corresponding to (021) and (004) crystal surface of α-Mg but increases the diffraction peak intensity of (112) crystal surface.

Such changes associated with the evolution of β-phase are ascribed to the acoustic streaming. During the solidification process, the Zn element is enriched on the grain boundaries usually, which makes the laminar $MgZn_2$ phase to form easily. When ultrasonic vibration is introduced in the melt, acoustic streaming accelerates

Positions	Element (at%)		
	Mg	Zn	Total
A	61.11	38.89	100
B	70.96	29.04	100
C	65.36	34.64	100

Table 2.
EDS results of positions in **Figure 14**.

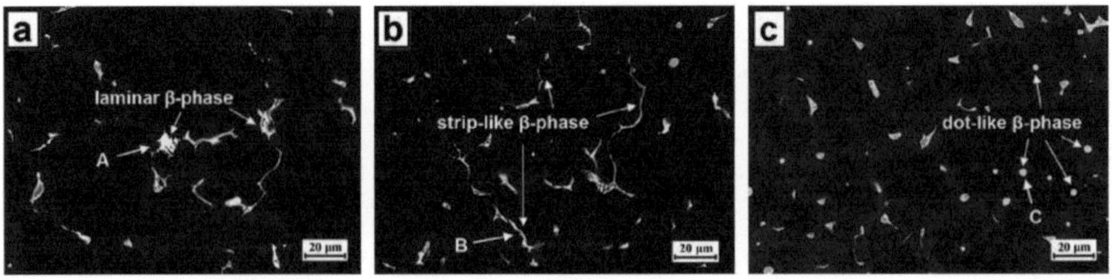

Figure 14.
SEM images of the β-phase in as-cast ZK60 alloys (a) without ultrasonic treatment; (b) with 1200 W SUF treatment; and (c) with 600 W DUF treatment.

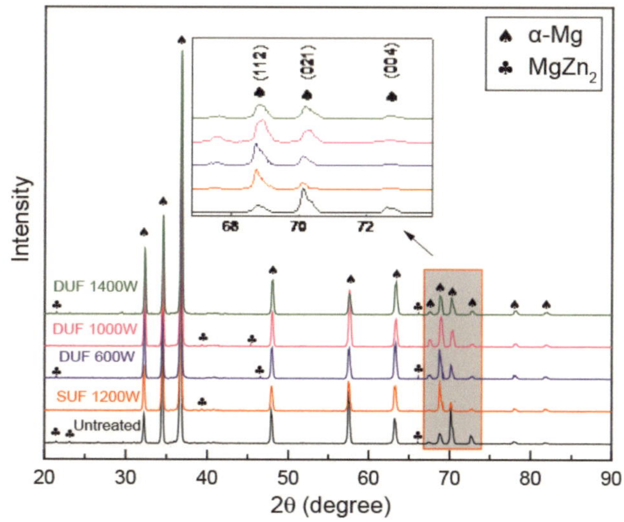

Figure 15.
XRD patterns of as-cast ZK60 alloys treated by different ultrasonic conditions.

the heat and mass transfer and then relieves the enrichment of Zn element. Thus, tiny and dot-like $MgZn_2$ phases are produced with dispersed distribution.

7. Ultrasonic melt treatment on DC casting process of magnesium billet

In the above sections, the ultrasonic treatment is employed to treat the magnesium melt in laboratory scale with small volume of melt. This section displays the application of ultrasonic treatment in direct-chill (DC) casting of magnesium billet. The DC casting method is the most popular way to produce large-sized magnesium billet [21]. However, due to the unique features of magnesium alloys, coarse grains, developed dendrites, wide columnar crystal area, and inhomogeneous structure appear in billet usually, which have adverse effect in subsequent processing. The traditional ultrasonic treatment technology generally has limited action area and then cannot satisfy the requirement of the treatment for large volume melt. Therefore, to overcome this weakness, we apply a unique ultrasonic treatment technology, namely, the variable-frequency ultrasonic treatment technology. In addition, the electromagnetic field is also a helpful external field for DC casting of magnesium alloys. In this section, we will show you the comparison of traditional fixed-frequency field (SUF), variable-frequency ultrasonic field (VUF), and low-frequency electromagnetic field (LEF) on DC casting of magnesium alloy. The interaction between LEF and VUF is also discussed [22–24].

Figure 16 demonstrates the experimental apparatus for ultrasonic DC casting. Three synergistic systems constitute the whole experimental setup, namely, the casting system, the low-frequency electromagnetic system, and the ultrasonic vibration system. The ultrasonic wave was provided by a self-designed ultrasonic vibration system, which comprised an ultrasonic generator, an ultrasonic transducer, a waveguide, and a 35 mm diameter stainless steel acoustic horn. The ultrasonic generator could produce two kinds of ultrasonic vibration fields, namely, the fixed-frequency ultrasonic field (FUF) and variable-frequency ultrasonic field (VUF). The frequency of FUF was fixed at 20 kHz, while the VUF had a 20 kHz center frequency and a 200 Hz changing frequency. The electromagnetic control unit and the induction coil provided the low-frequency electromagnetic field (LEF). The frequency of LEF was 20 Hz and the alternating current was 150 A. A resistance

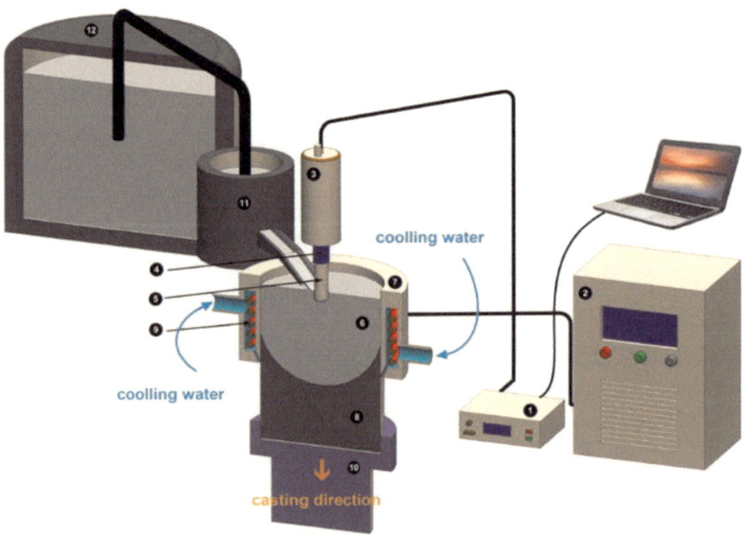

Figure 16.
Schematic image of the experimental apparatus: (1) ultrasonic power supply, (2) electromagnetic field control unit, (3) ultrasonic transducer, (4) acoustic waveguide, (5) acoustic radiator/horn, (6) magnesium alloy melt, (7) crystallizer, (8) billet, (9) magnetic coil, (10) dummy bar, (11) tundish, and (12) smelting furnace.

furnace, a crystallizer with the Ø255 mm aluminum ring, a tundish, and a casting control unit compose the casting system.

At first, the commercial pure magnesium, pure aluminum, pure zinc, and anhydrous MnCl2 were melted in a resistance furnace at 740°C with the protection of CO2 + 0.5% SF6 (1:2) atmosphere. As shown in **Figure 16**, the melt was transferred to the crystallizer at the temperature of 670°C from the crucible. With the downward movement of the casting machine, the AZ80 billets (Ø255 mm) were made. During the casting process, the preheated ultrasonic horn was inserted 50 mm under the interface of the melting liquid in order to introduce the ultrasonic vibration into the melt. In the same time, the LEF worked together to input the electromagnetic field. Samples for microstructure characterization were ground by the different grades of SiC papers and then polished and etched using a solution of 4.2 g picric acid to reveal their microstructures. The characterizations of microstructure were obtained using an optical microscopy (OM) and a scanning electron microscopy (SEM) with energy-dispersive spectroscopy (EDS).

Figure 17 displays the microstructure of as-cast AZ80 billet casted with different external field treatments. Obviously, a wide columnar crystal zone is observed in the center of billet with approximate 22.2 mm length. The α-Mg grains are coarse with large grain size and high degree of inhomogeneity. The FUF treatment refines the grains located in the center area but has little effect on the edge of billet. The grain refinement efficiency decreases from center to edge of billet. It is interesting to find that the columnar crystal area disappears. The LEF also has ability to refine α-Mg grains, while its refinement tendency is different from the FUF, showing increase trend from center to edge of billet. The LEF can only reduce the width of columnar crystal area to columnar crystal area to 12.8 mm. The VUF treatment shows the best grain refinement capacity. All the grains along center to edge become very small with global form. In addition, grains have homogeneous distribution in whole billet. With VUF and LEF treatment, the grain is also refined dramatically with the enhancement of homogeneity as well. It notes that the LEF + VUF may further promote the refinement efficiency.

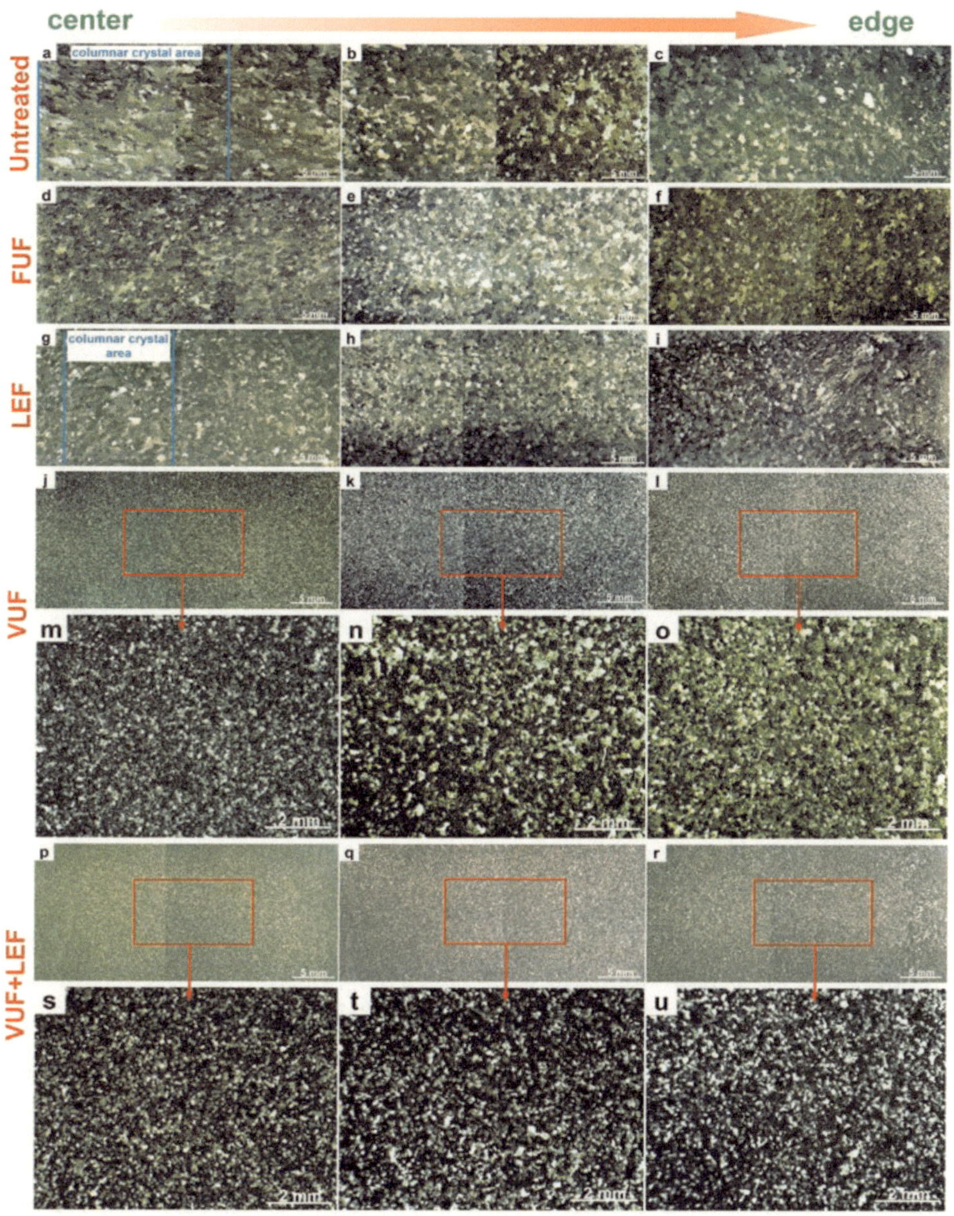

Figure 17.
Microstructure from center to edge of as-cast AZ80 billet with different external field treatments: (a)–(c) without external field; (d)–(f) with FUF treatment; (g)–(i) with LEF treatment; (j)–(l) with VUF treatment; (p)–(r) with VUF + LEF treatment; (m)–(o) and (s)–(u) are partial enlarged picture.

Figure 18 reflects the statistic of grain size and its distribution of as-cast AZ80 treated by different external fields. The grain size of untreated billet is 679–1454 μm which reduces to 752–916 μm with FUF treatment. The value associated with the LEF treatment is lower than FUF showing a figure of 732–1027 μm. The VUF displays the excellent refinement efficiency, with the grain size of 150–241 μm. This value is further reduced to 116–141 μm after VUF + LEF treatment.

The homogeneity of grain size is also very important particularly for large-sized billet. Therefore, to evaluate this item, the linear fitting and standard deviation (SD) of grain size distribution are employed. The slope of fitted line reflects the

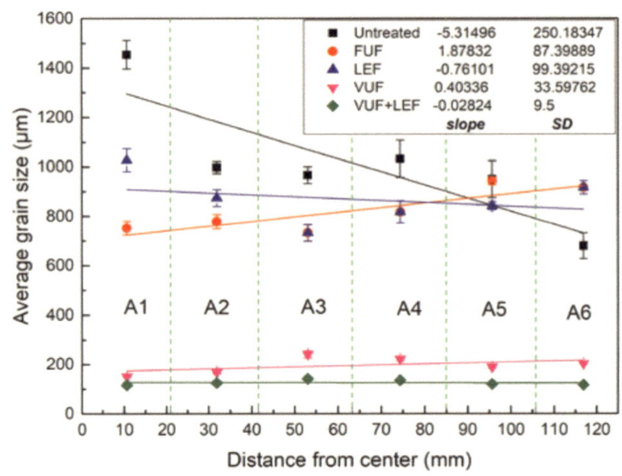

Figure 18.
Grain size distribution of as-cast AZ80 under different external field treatments.

characteristics of grain distribution, and the SD value shows the homogeneity. In case of untreated billet, a negative slope (−5.314) is obtained with the large SD value (250.183), suggesting that the grain size reduces from center to edge and bad homogeneity. After FUF treatment, the slope becomes positive to 1.878, and the SD decreases to 87.399. These results agree with the microstructure observation in **Figure 17**, implying the better refinement caused by FUF in the center part of billet and improvement of homogeneity. The LEF has negative slope and relatively small SD value of −0.761 and 99.392, respectively, which means better grain refinement efficiency in the edge area of billet. The VUF-treated billet has the negative slope similar with the FUF, while the slope is rapidly reduced to 0.403, with the small SD value of 33.598 as well, suggesting the good homogeneity. As for the VUF + LEF-treated billet, the slope and SD are further reduced to −0.028 and 9.5, respectively. Thus, the LEF can help promote the homogeneity.

To better explain and reflect the mechanisms of external field treatment on microstructure evolution of AZ80 alloy, schematic diagrams are given in in **Figure 19**. As mentioned above, the cavitation-enhanced heterogeneous nucleation and dendrite fragmentation are two main reasons to refine Mg grains. As shown in **Figure 19(b)**, the ultrasonic cavitation bubble is produced and enlarged because of the negative acoustic pressure, which absorbs heat from melt and generates the local super cooling on the bubble's surface, leading to heterogeneous nucleation on the surface of bubble. The enlarged bubble shrinks and collapses due to the positive acoustic pressure and releases huge impact force to melt, which can fragment the initial crystal nuclei and dendrites. This process is also an effective method to increase the number of nuclei in melt. What is more, it can remove the columnar crystals, and that is why the columnar crystal area disappears in **Figure 17(d)**. These two simultaneously worked mechanisms considerably increase the number of nuclei of α-Mg. Then in the acoustic streaming, the return circulated macroscopic flow in melt (shown in **Figure 19(b)**) carries the nuclei to wider areas, showing the decrease of absolute value of slope and SD compared with the untreated billet. The linear fitting results note that FUF-treated billet has the positive slope of grain size distribution. The sound attenuation should respond for this phenomenon. The acoustic intensity I is inversely related to the propagation distance x, as shown in Eq. (7):

$$I = I_0 e^{-2\alpha x} \tag{7}$$

where α the attenuation coefficient in melt and x the propagation distance. Because of the high attenuation coefficient α value in magnesium melt, the acoustic intensity becomes weak in the edge area of billet, reducing the grain refinement efficiency.

According to the experimental results, the VUF has the better grain refinement ability than FUF. To understand the reason behind this phenomenon, a simple numerical simulation for acoustic pressure distribution in melt is calculated by finite element method. The acoustic pressure distribution is calculated by Eq. (8):

$$\frac{1}{\rho_0 c_0^2}\frac{\partial^2 p}{\partial t^2} - \nabla \cdot \left(\frac{1}{\rho_0}(\nabla p)\right) = 0 \qquad (8)$$

where p is the solved acoustic pressure, t is the time, and ρ_0 and c_0 are the melt's density and sound speed in magnesium melt. In this simulation, the solidified Mg billet is not considered. Thus, the physical model only shows the Mg melt during the casting process based on the experimental data. Four boundary conditions are as follows. The melt-air surface is set as the soft wall condition where $P = 0$. The side wall of ultrasonic horn is set as the hard wall condition, where the acoustic wave is totally reflected. The tip surface of ultrasonic horn is set as the acoustic pressure condition, where $p = p_0 \sin(2\pi t(f_0 + \sin(2\pi f_1 t)))$ represents the variable-frequency ultrasonic vibration mode, and the p_0, f_0, and f_1 are the amplitude of acoustic pressure, center frequency, and the changing frequency, respectively. The surface of ring and mushy zone are set as the acoustic impedance where the acoustic wave can be absorbed. **Table 3** lists the parameters used in this simulation.

Figure 20 displays the acoustic pressure distribution and corresponding cavitation area. In case of FUF, the maximum negative acoustic pressure is located on the surface of ultrasonic horn with the value of −2.88 MPa. As for VUF, the maximum negative acoustic pressure appears in the area below the tip of horn with the value of −4.58 MPa. Based on the research results of Eskin, the cavitation threshold of light materials is about 1.1 MPa. The results show that the cavitation area of FUF is mainly located in a small area close to the ultrasonic horn but the VUF has larger and wider cavitation area. Therefore, the VUF treatment can both improve the maximum acoustic pressure and the cavitation area. On the one hand, the wider cavitation area can produce more cavitation bubble. On the other hand, higher ultrasonic pressure can bring larger cavitation bubble and promote the acoustic streaming effect. Bigger cavitation bubble means that there is more heat absorbed from the melt which then generates greater degree of supercooling, resulting in increase of number of nuclei, as shown in **Figure 19(c)**. Larger cavitation bubble can also promote the destructive power of impact force released by the collapse of cavitation bubble, improving the dendrite fragmentation effect. The impact force (F_{impact}) is given as [25].

Parameter	Description	Value
p_0	Amplitude acoustic pressure	3.22×10^6
f_0	Center frequency	20 kHz
f_1	Changing frequency	200 Hz
c	Sound speed	4000 m/s
ρ	Density of melt	1780 kg/m^3
Z	Acoustic impedance	4.2×10^7 Pa•s/m

Table 3.
Parameters used in numerical simulation.

$$F_{impact} = I_{in} \cdot S \cdot e^{-2\alpha x} \cdot \frac{e_1}{c} \tag{9}$$

where I_{in} is the input acoustic intensity outside the bubble; S is the surface of cavitation bubble; α is the attenuation coefficient; x is the transmitted position; e_1 is a unit vector ultrasonic wave associated with the propagation direction; and c is the speed of ultrasound in the magnesium melt. Therefore, the impact force (F_{impact}) has strong relation with the S and I_{in}. On the one hand, the I_{in} is positively relative with the acoustic pressure, according to $I_{in} = p^2/2\rho c$. Thus, the higher acoustic pressure, the larger is I_{in}, suggesting that the VUF can enhance the impact force. On the other hand, higher acoustic pressure also produces the large-sized cavitation bubble, namely, the enlargement of S, leading to the enhancement of impact force (F_{impact}). It is also noticed that the VUF makes the grains more uniform. The improved acoustic streaming effect is responsible for this phenomenon. Due to the fact that the acoustic pressure gradient holds the key to the intensity of acoustic streaming, the higher maximum ultrasonic pressure of VUF can dramatically improve the action sphere of acoustic streaming. Stronger acoustic has better transportation capability and can carry nuclei to further place, promising the homogeneity of billet.

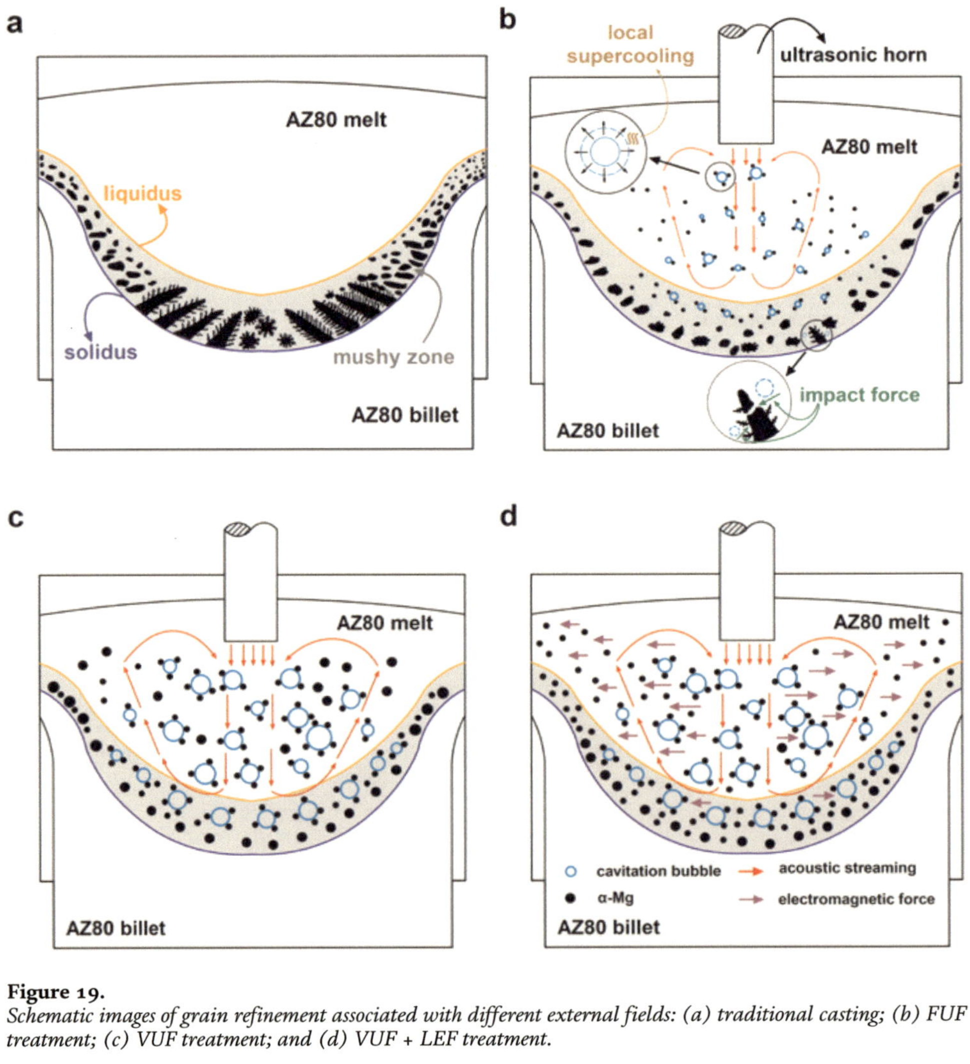

Figure 19.
Schematic images of grain refinement associated with different external fields: (a) traditional casting; (b) FUF treatment; (c) VUF treatment; and (d) VUF + LEF treatment.

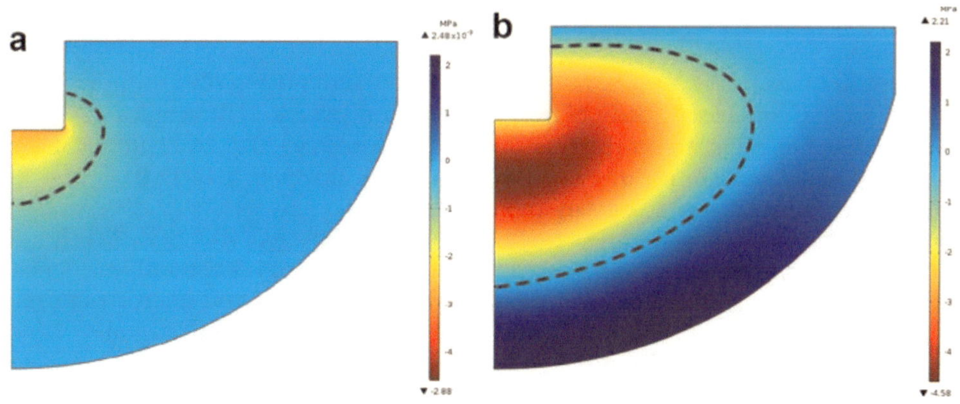

Figure 20.
Acoustic pressure distribution and relevant cavitation area in magnesium melt (a) FUF and (b) VUF.

According to the experimental results, the LEF can also refine the microstructure of AZ80 alloy and improve its homogeneity. In this case, the Lorentz force produced by the electromagnetic field introduces the stirring effect to the melt, which enforces the melt moving from center to edge along the transverse direction. Consequently, the melt with high temperature in center part of melt is carried to the cold mold wall, accelerating the heat and mass transfer. The temperature field becomes uniform, and temperature gradient is reduced as well, which helps in maintaining the nuclei. In addition, the vibrating forces can also break the initial solidified grains formed on the mold wall and then refine the grain.

Now, it is worth to discuss the mechanisms of combination application of VUF and LEF, because it shows the best grain refinement efficiency, with the grain size of 116–141 μm, slope of −0.028, and SD of 9.5. These two fields have interaction, as show in **Figure 19(d)**. According to the discussion above, the main effect of two fields is different. As for the VUF, increasing the number of nuclei in melt is its uppermost mission. The acoustic streaming is its secondary effect. In case of LEF, accelerating the mass and het transportation is the main effect because it can create a comfortable environment for nuclei. Thus, the VUF generates the large number of nuclei and LEF charge to transport them and make sure they can grow. Finally, the grains become fine with homogeneous distribution.

Author details

Xingrui Chen and Qichi Le*
Key Lab of Electromagnetic Processing of Materials, Ministry of Education, Northeastern University, Shenyang, People's Republic of China

*Address all correspondence to: qichil@mail.neu.edu.cn

References

[1] Chen X, Liao Q, Niu Y, Jia Y, Le Q, Ning S, et al. Comparison study of hot deformation behavior and processing map of AZ80 magnesium alloy casted with and without ultrasonic vibration. Journal of Alloys and Compounds. 2019; **803**:585-596. DOI: 10.1016/j.jallcom.2019.06.242

[2] Xu T, Yang Y, Peng X, Song J, Pan F. Overview of advancement and development trend on magnesium alloy. Journal of Magnesium and Alloys. September 2019;**7**(3):536-544. DOI: 10.1016/j.jma.2019.08.001

[3] Chu C, Hu Z, Li X, Yan H, Wu X, Mai Y. Evolution and distribution of Al2Sm phase in as-extruded AZ61–xSm magnesium alloys during semi-solid isothermal heat-treatment. Transactions of the Nonferrous Metals Society of China. 2018;**28**:1311-1320. DOI: 10.1016/S1003-6326(18)64768-5

[4] Kula A, Tokarski T, Niewczas M. Comparative studies on the structure and properties of rapidly solidified and conventionally cast AM60 magnesium alloys. Materials Science and Engineering A. 2019;**759**:346-356. DOI: 10.1016/j.msea.2019.05.044

[5] Plesset MS, Prosperetti A. Bubble dynamics and cavitation. Annual Review of Fluid Mechanics. 1977;**9**: 145-185

[6] Anderson PW. Absence of diffusion in certain random lattices. Physics Review. 1958;**109**:1492-1505. DOI: 10.1103/PhysRev.109.1492

[7] Liu X, Zhang J, Li H, Le Q, Zhang Z, Hu W, et al. Electrical resistivity behaviors of liquid Pb–Sn binary alloy in the presence of ultrasonic field. Ultrasonics. 2015;**55**:6-9. DOI: 10.1016/j.ultras.2014.07.008

[8] Liu X, Zhang Z, Hu W, Le Q, Bao L, Cui J, et al. Study on hydrogen removal of AZ91 alloys using ultrasonic argon degassing process. Ultrasonics Sonochemistry. 2015;**26**:73-80. DOI: 10.1016/j.ultsonch.2014.12.015

[9] Liu X, Zhang C, Zhang Z, Xue J, Le Q. The role of ultrasound in hydrogen removal and microstructure refinement by ultrasonic argon degassing process. Ultrasonics Sonochemistry. 2017;**38**:455-462. DOI: 10.1016/j.ultsonch.2017.03.041

[10] Liu X, Xue J, Zhao Q, Le Q, Zhang Z. Effects of radiator shapes on the bubble diving and dispersion of ultrasonic argon process. Ultrasonics Sonochemistry. 2018;**41**:600-607. DOI: 10.1016/j.ultsonch.2017.10.026

[11] Sigworth GK, Engh TA. Chemical and kinetic factors related to hydrogen removal from aluminum[J]. Metallurgical Transactions B. 1982;**13** (3):447-460

[12] Papamantellos D, Lange KW, Okohira K, Schenck H. A mathematical approach for the mass transfer between liquid steel and an ascending bubble. Metallurgical and Materials Transactions B: Process Metallurgy and Materials Processing Science. 1971;**2**: 3135-3144. DOI: 10.1007/BF02814966

[13] Zhang ZQ, Le QC, Cui JZ. Effect of Ultrasonic Treatment on Microstructures of Mg-Ca Binary Alloy [C]//Advanced Materials Research. Trans Tech Publications; 2010;**139**: 677-680

[14] Shao Z, Le Q, Zhang Z, et al. Effect of ultrasonic power on grain refinement and purification processing of AZ80 alloy by ultrasonic treatment[J]. Metals and Materials International. 2012;**18**(2): 209-215

[15] Eskin GI. Principles of ultrasonic treatment: Application for light alloys

melts. Advanced Performance Materials. 1997;**4**:223-232. DOI: 10.1023/A:1008603815525.

[16] Hunt JD, Jackson KA. Nucleation of solid in an undercooled liquid by cavitation. Journal of Applied Physics. 1966;**37**:254-257. DOI: 10.1063/1.1707821

[17] Jin J, Ma H, Wang K, Yagoub AE-GA, Owusu J, Qu W, et al. Effects of multi-frequency power ultrasound on the enzymolysis and structural characteristics of corn gluten meal. Ultrasonics Sonochemistry. 2015;**24**:55-64. DOI: 10.1016/j.ultsonch.2014.12.013

[18] Shankar PM, Chapelon JY, Newhouse VL. Fluid pressure measurement using bubbles insonified by two frequencies. Ultrasonics. 1986;**24**:333-336. DOI: 10.1016/0041-624X(86)90004-1

[19] Phelps AD, Ramble DG, Leighton TG. The use of a combination frequency technique to measure the surf zone bubble population[J]. The Journal of the Acoustical Society of America. 1997;**101**(4):1981-1989

[20] Chen X, Ning F, Hou J, Le Q, Tang Y. Dual-frequency ultrasonic treatment on microstructure and mechanical properties of ZK60 magnesium alloy. Ultrasonics Sonochemistry. 2018;**40**:433-441. DOI: 10.1016/j.ultsonch.2017.07.027

[21] Jia Y, Chen X, Le Q, Wang H, Jia W. Numerical study on action of HMF, PMF, DHMF, and DPMF on molten metal during electromagnetic casting. International Journal of Advanced Manufacturing Technology. 2019;**103**:201-217. DOI: 10.1007/s00170-019-03501-y

[22] Chen X, Le Q, Wang X, Liao Q, Chu C. Variable-frequency ultrasonic treatment on microstructure and mechanical properties of ZK60 alloy during large diameter semi-continuous casting. Metals. 2017;7:173. DOI: 10.3390/met7050173

[23] Chen X, Jia Y, Liao Q, Jia W, Le Q, Ning S, et al. The simultaneous application of variable frequency ultrasonic and low frequency electromagnetic fields in semi continuous casting of AZ80 magnesium alloy. Journal of Alloys and Compounds. 2019;**774**:710-720. DOI: 10.1016/j.jallcom.2018.09.300

[24] Jia Y, Wang H, Le Q. Transient coupling simulation of multi-physical field during pulse electromagnetic direct-chill casting of AZ80 magnesium alloy. International Journal of Heat and Mass Transfer. 2019;**143**:118524. DOI: 10.1016/j.ijheatmasstransfer.2019.118524

[25] Liu Q, Zhai Q, Qi F, Zhang Y. Effects of power ultrasonic treatment on microstructure and mechanical properties of T10 steel. Materials Letters. 2007;**61**:2422-2425. DOI: 10.1016/j.matlet.2006.09.027